Microfluidics and Lab-on-a-chip

Microfluidics and Lab-on-a-chip

Editor: Tami Clawson

MURPHY & MOORE
www.murphy-moorepublishing.com

www.murphy-moorepublishing.com

⊕ MURPHY & MOORE

Cataloging-in-publication Data

Microfluidics and lab-on-a-chip / edited by Tami Clawson.
   p. cm.
Includes bibliographical references and index.
ISBN 978-1-63987-702-7
1. Microfluidics. 2. Labs on a chip. 3. Nanofluids. 4. Chemical apparatus.
5. Microelectromechanical systems. I. Clawson, Tami.
TJ8534.M53 M53 2023
620.106--dc23

Murphy & Moore Publishing
1 Rockefeller Plaza,
New York City,
NY 10020, USA

ISBN 978-1-63987-702-7

Contents

Preface

Microfluidics is the science that studies the behavior of fluids through microchannels. Microfluidic technology is involved in manufacturing systems that manipulate small amounts of fluidics using micro-channels. It deals with the behavior, precise control, and manipulation of fluids that are constrained in small spaces in the range of sub-millimeters, where the surface forces dominate volumetric forces. It plays an important role in the design of systems which process low volumes of fluids to achieve multiplexing, automation, and high-throughput screening. It is used in the development of lab-on-a-chip (LOC) technology, DNA chips, micro-propulsion, inkjet print heads, and micro-thermal technologies. LOC devices are a part of microelectromechanical systems devices performing analyses on micro-amounts of fluids. They integrate and automate multiple high-resolution laboratory techniques such as synthesis and analysis of chemicals or fluid testing into a system that fits on a chip. This book outlines the processes and applications of microfluidics and lab-on-a-chip in detail. While understanding the long-term perspectives of the topics, the book makes an effort to highlight their impact as a modern tool for the growth of the discipline.

The information contained in this book is the result of intensive hard work done by researchers in this field. All due efforts have been made to make this book serve as a complete guiding source for students and researchers. The topics in this book have been comprehensively explained to help readers understand the growing trends in the field.

I would like to thank the entire group of writers who made sincere efforts in this book and my family who supported me in my efforts of working on this book. I take this opportunity to thank all those who have been a guiding force throughout my life.

Editor

Magnetic Field-Based Technologies for Lab-on-a-Chip Applications

Veronica Iacovacci, Gioia Lucarini,
Leonardo Ricotti and Arianna Menciassi

Abstract

In the last decades, LOC technologies have represented a real breakthrough in the field of in vitro biochemical and biological analyses. However, the integration of really complex functions in a limited space results extremely challenging and proper working principles should be identified. In this sense, magnetic fields revealed to be extremely promising. Thanks to the exploitation of external magnetic sources and to the integration of magnetic materials, mainly high aspect ratio micro-/nanoparticles, non-contact manipulation of biological and chemical samples can be enabled. In this chapter, magnetic field-based technologies, their basic theory, and main applications in LOC scenario will be described by foreseeing also a deeper interaction/integration with the typical technologies of microrobotics. Attention will be focused on magnetic separation and manipulation, by taking examples coming from traditional LOC devices and from microrobotics.

Keywords: Lab-on-chip, microrobotics, magnetic nanoparticles, magnetophoresis, magnetic manipulation

1. Introduction

The need of reliable, precise, and fast techniques for biochemical and biological analysis has fostered the search for miniaturized systems integrating multiple laboratory techniques, assays, and procedures into a really small chip, up to few square centimeters in size. These small platforms, named lab-on-a-chip (LOC) or, less frequently, micro total analysis systems (μTAS), have historically been fabricated in silicon and/or glass using semiconductor processing techniques. More recently, polymer-based devices emerged, thanks to the introduction of soft lithography [1]. LOC devices can be considered as sophisticate microsystems embedding mechanical, electronic, and fluidic functions [2], aiming at mixing, pumping, and manipulating samples. It is possible to identify a wide literature concerning LOC systems, in which a variety of applications ranging from biological assays, drug sorting and testing, DNA extraction, cell manipulation, etc., have been explored.

The use of LOC devices for laboratory tasks execution offers several advantages: reduced sample and reagent volumes, fast sample processing, high sensitivity and spatial resolution, increased detection accuracy, low contamination, high throughput, and reliability thanks to the possibility to automate some processes, without depending on human operator skills [3].

Due to the really small dimensions of LOC devices, the major role is played by surface effects with respect to inertial ones. Consequently, traditional actuation strategies cannot be exploited for actuation in LOCs. Fluidic actuation is the most commonly employed strategy, but electrostatic, magnetic, and chemical motion has been reported as well.

LOC systems can be distinguished in continuous-flow and stationary devices, depending on the role played by the fluidic actuation in the execution of the desired tasks. In continuous-flow devices, microfluidic forces are responsible for the effects experienced by the sample (e.g., beads, liquids, or droplets). In static flow devices, although the working environment is still a fluid, additional actuation strategies, such those based on magnetic fields, are exploited for effectively executing the desired task.

Depending on the working environment and on the object lengthscale, the most effective physical principle to be exploited in order to achieve the desired objective can change significantly. **Figure 1** shows the trend of different physical effects at varying of the manipulated object dimension. At the microscale, due to the capillary forces and to low Reynolds numbers, it is quite hard to manipulate or mix liquids and particles by exploiting only fluidic forces or direct manipulation, and the exploitation of other actuation strategies showing high efficiency at the microscale is required. In this sense, magnetic field-based strategies exploitation could be a valid solution. In LOC scenarios, in fact, the magnetic field sources can be really close to the working environment, thus compensating the rapid decay of magnetic force with the distance between the source and the object [4]. Furthermore, the exploitation of magnetic fields enables non-contact manipulation [5] also for biological samples, thus paving the way for a wide variety of applications in biology and medicine. In this chapter, the force balance acting on a micro-object in a LOC will be analyzed with a particular focus on magnetism basic theory. The exploitation of magnetic fields for torque and force generation will be considered, especially for magnetic separation and magnetic manipulation applications. Techniques employed both to endow an object with magnetic properties and to characterize it will be described. Finally, potential applications of magnetic field-based strategies in LOCs will be reviewed. Throughout the chapter, technologies and examples not typical of LOCs but deriving, for example, from the world of microrobotics will be introduced, thus foreseeing a deeper and deeper interaction/integration between these two fields.

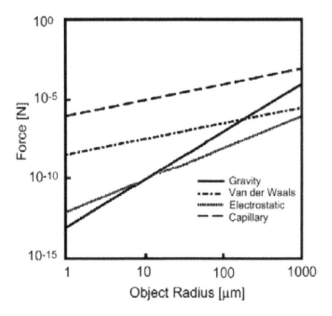

Figure 1. Scaling of different forces in function of the size of the object.

2. Physics at the microscale: key principles

According to Newton's second law, when considering a magnetic microcarrier with mass m_p and moving in a fluidic environment with velocity v, the following equilibrium equation should be considered:

$$m_p \frac{dv}{dt} = F_m + F_g + F_d + F_b \tag{1}$$

Several physical effects, including the magnetic force F_m, the fluidic drag force F_d, the net gravitational force F_g that take into account also the fluidic lift effect, and the Brownian interaction force F_b contribute to the force balance of the moving object. In the following, the Brownian force, representing fluid-object and inter-objects interactions, will be neglected as it is really weak with respect to the other contributions. The other forces contributing to the equilibrium will be analyzed more in detail.

2.1. Fluidic drag force

Navier–Stokes equations completely define a fluid velocity in space and time. From these equations, it is possible to derive the Reynolds number (Re), a dimensionless quantity that is proportional to the ratio between the fluid's inertia and its viscosity and that allows to define a fluid's behavior when it flows around an object. Given the fluid density ρ, the dynamic viscosity η, the object maximum velocity with respect to the fluid v and a characteristic linear dimension L, Re can be defined as:

$$Re = \frac{\rho v L}{\eta} \tag{2}$$

Usually, in both microrobotics and LOC applications, a low-Re regime, typically defined for $Re < 10$, applies. At low Re, surface and capillary forces play an important role compared to inertia and temporal variations of the flow pattern. Due to the relative importance of surface effects, flow at low Re-number strongly depends on object geometry. Thus, it is interesting to derive the viscous drag force acting on the object. By approximating the object to a sphere with radius r put in an infinite extent of fluid, the viscous drag force can be calculated as a linear function of the sphere's velocity through the fluid:

$$F_d = 6\pi\eta r(v_f - v_p) \tag{3}$$

where v_f and v_p are the fluid and the sphere velocity, respectively.

2.2. Gravitational force

Inertial and gravitational forces play a minor role at low Re. When considering a micro-object immersed in a liquid, usually net gravitational force is taken into account. In fact upthrust forces, responsible for body buoyancy, should be considered in addition to gravitational force, which acts in the opposite direction:

$$F_g = -V_p\left(\rho_p - \rho_f\right)g \tag{4}$$

In Eq. (4), V_p and ρ_p are object volume and density, ρ_f is the fluid density, and g is the gravity acceleration.

2.3. Magnetic force

The force acting on an object immersed in a magnetic field depends both on the field features and on the object properties. The magnetic force acting on a magnetic microstructure can be modeled by using the "effective" dipole moment method, in which a magnetic object is replaced by an "equivalent" point dipole with a moment $m_{p,eff}$ [6]. The force on the dipole is given by:

$$F_m = \mu_f(m_{p,eff} \cdot \nabla)B \tag{5}$$

where μ_f is the magnetic permeability of the medium, $m_{p,eff}$ is the "effective" dipole moment of the object, and B is the magnetic field produced by an external source acting at the center of the target object, where the equivalent point dipole is located.

The dipole moment m strictly depends on object volume and magnetic properties and it can be defined as:

$$m = M \cdot V \tag{6}$$

where M and V are the magnetization and the volume of the dipole, respectively.

As mentioned, the force exerted on such dipole varies upon the features of the magnetic field sources. It also depends on the distance between the source and the target object. If considering a permanent magnet, the magnetic field density at a generic point P can be expressed as:

$$B = \frac{\mu_0}{4\pi}\left(\frac{3(m \cdot r)r}{|r|^5} - \frac{m}{|r|^3}\right) \tag{7}$$

with r being the distance vector connecting the field source and the point P.

3. Classification of magnetic materials, magnetization, and characterization of micro-objects

One of the greatest advantages of magnetic actuation lies in the possibility to transfer powering and actuation in a wireless fashion. Remote magnetic actuation relies on the coupling, namely the creation and maintenance of a magnetic link, between two objects showing magnetic properties. Typically, an external control platform, based on permanent magnets, electromagnets, or a combination of them, and a micro-object, that could be a magnetic bead, a magnetized cell, or a microrobot, constitute the key elements. Materials behavior in response to a magnetic field depends on the material atomic organization. In particular, the spatial organization of the material microscopic domains and the possible changes in this organization produced by the presence of an external magnetic field determine the material response. Indeed, the magnetization induced in a material is proportional to the ability of these domains to align or to form cooperative structures when a magnetic field is applied. This ability can be described by means of the magnetic susceptibility χ, a non-dimensional parameter defined by the ratio of the magnetization M induced in the material and the applied magnetic field H.

$$M = \chi H \tag{8}$$

Depending on this parameter, it is possible to classify magnetic materials in three main categories: diamagnetic, paramagnetic, and ferromagnetic (**Figure 2**) [7]. Diamagnetic materials, such as bismuth or brass, have no net atomic or molecular magnetic moment and do not retain magnetization when the external magnetic field is removed. When these materials are subjected to an applied field, atomic currents generate and produce a bulk magnetization antiparallel to the field H, thus resulting in negative and negligible susceptibility χ levels (~10^{-6} to ~10^{-3}). Paramagnetic materials have a net magnetic moment at the atomic level which shows a random orientation when no magnetic field is acting. When the magnetic field H is applied, the moment tends to align with it. The susceptibility of such materials is in the range 10^{-6}–10^{-1}. Ferromagnetic materials, such as iron, nickel, and cobalt, on the other hand, have a net magnetic moment at the atomic level, but unlike paramagnetic materials, they show a strong coupling between neighboring moments as they align all in the same direction and parallel to each other to produce a larger magnetization state. This coupling gives rise to a spontaneous alignment of the moments over macroscopic regions, called domains, which undergo further alignment when the material is subjected to an applied field. Ferromagnetic materials can be permanently magnetized since they are able to retain residual magnetization after the removal of the applied magnetic field. They can be furtherly classified as soft or hard materials: The first ones are featured by a high permeability and a low coercivity H_c (the coercivity is defined as the magnetic field intensity needed to reduce the magnetization of a ferromagnetic material from its complete saturation to zero). This makes them easy to be magnetized and demagnetized. The second ones have a relatively low permeability and high coercivity which make them more suitable for the fabrication of permanent magnets [8, 9].

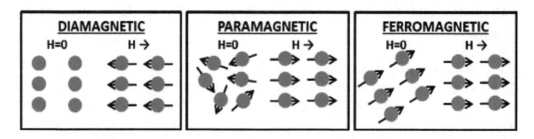

Figure 2. Schematic representation of diamagnetic, paramagnetic, and ferromagnetic materials microscopic structures at rest and in the presence of a magnetic field H.

To enable magnetic field-mediated task execution in a LOC, it is necessary to provide the objects to be manipulated with magnetic properties. In some cases, magnetic manipulation relies on the intrinsic magnetic properties of the sample, as in the case of red blood cells [10]. More frequently, labeling and internalization of magnetic material, or fabrication of magnetic microcarriers, are required. To this aim, magnetic micro- and nanoparticles have gained growing attention in LOC systems and in microrobotics in general. Usually, polymeric or silica microparticles with embedded iron oxide nanoparticles are used. Simple iron oxide nanoparticles are also used, mainly magnetite (Fe_3O_4) and maghemite (γ-Fe_2O_3) ones. Due to the reduced dimensions of the magnetic core (diameter <1 μm), these particles usually consist of single magnetic domains showing a superparamagnetic behavior. The main advantages of using magnetic particles are that they have a large surface-to-volume ratio; they can be conveniently biofunctionalized, thus favoring their coating or enabling labeling molecules. To provide a micro-object with magnetic properties, two main strategies can be employed: (1) labeling with magnetic particles or (2) particle internalization. In both cases, sample incubation in the presence of a relatively high concentration of particles is required.

Sample magnetic labeling relies on the possibility to properly functionalize particle surface to enable the binding with functional groups exposed on sample surface. This applies, for example, for cell labeling: functional micro- or nanoparticles are conjugated with antibodies corresponding to specific cell surface antigens [11].

In the case of internalization, the magnetic particles are included in the sample structure itself by embedding the magnetic material during the micro-object fabrication process or exploiting transfection and magnetofection techniques in the case of biological samples. In this case, superparamagnetic iron oxide nanoparticles (SPIONs), usually properly modified to promote internalization, for example, through polystyrene or dextran coatings, or exploiting other transfection agents, such as peptides or antibodies [12], are widely employed.

An alternative to sample magnetization through labeling or internalization is the exploitation of magnetic carriers or manipulation systems that avoid a direct contact between sample and magnetic material.

In this case, magnetic properties can be imparted to a carrier, by simply including magnetic materials in its structure. To this aim, not only SPIONs have been employed: the integration of ferromagnetic materials, for example, in the form of powder, has been investigated in applications in which high magnetic responsivity and residual magnetization were required. Ferromagnetic materials, such as Ni, have been employed also in the form of surface coating, obtained through sputtering or evaporation techniques, with the aim to provide micro-objects showing complex geometries, fabricated, for example, through 2D or 3D lithography techniques, with magnetic properties [13].

Once identified the methods allowing to magnetize the samples to be manipulated, it can be useful to briefly describe some techniques allowing to properly characterize a magnetic micro-object. When designing the hardware for a magnetic manipulation or separation platform, it is useful (in some cases even mandatory) to precisely know the magnetic properties of the beads or structures to be manipulated. Particularly, interesting parameters, most commonly evaluated, are the magnetic susceptibility χ, the saturation magnetization M_s and the coercivity field H_c. Considering that small entities usually show weak magnetic properties, traditional technologies employed at the macroscale, such as Hall sensor-based probes, do not result suitable for their characterization. Microstructured magnetic materials can be properly characterized through both inductive- and force-based techniques. Inductive methods, such as the vibrating sample magnetometry (VSM) and the superconducting quantum interference device (SQUID) magnetometry, are usually employed for magnetic characterization at the micro-/nanoscale. In both cases, the measurement can be carried out at variable temperature and by applying different magnetic fields, thus allowing to obtain the typical material magnetization curves in a specific range of temperatures. In VSM, a magnetic sample is vibrated within a uniform magnetic field: sample vibration induces a current in dedicated sensing coils; by measuring the resulting voltage induced into the coil, it is possible to obtain sample magnetic moment and to magnetically characterize it. The sensitivity of this kind of technique can reach 10^{-6} emu. When the samples are really diluted or show really weak magnetic properties, thus claiming for higher sensitivities, SQUID-based magnetometry can be a suitable solution, enabling to reach sensitivities up to 10^{-8}–10^{-12} emu. The magnetic properties of the material are measured by detecting quantum mechanical effects in conjunction with superconducting detection coils. In both VSM and SQUID magnetometry, however, the duration of a single measurement is in the order of some hours. This obviously represents a strong limitation for all cases in which the characterization of a large number of samples is needed. On the other hand, force-based methods, such as Gouy and Faraday balances, rely on the change in weight of a magnetic material when it is subjected to a uniform magnetic field. Commercial systems based on the Faraday method, such as the alternating gradient magnetometer (AGM), provide sensitivities in the 10^{-8}–10^{-9} emu range with really fast measurement procedures [14, 15].

4. Magnetic actuation in LOC: principles and exploited hardware

Some applications of magnetic fields in LOCs have already been mentioned and range from magnetic separation for chemical and biological analyses to sample manipulation for drug screening and cell sorting. In terms of magnetic actuation of samples, it is possible to classify the various applications in two main categories: (1) magnetic separation and (2) magnetic manipulation. In the first case, two or more classes of objects are separated depending on their magnetic properties, but without any need to properly drive them along complex paths or to guarantee the execution of specific tasks; in this case, magnetic fields are responsible for separation, but transportation is usually provided by fluidic forces. In the second case, a more accurate control is required to enable a single magnetic object or a swarm of them to follow a planned trajectory or to perform a specific task; larger magnetic fields and forces are required in this case, as they are responsible also for object transport.

4.1. Magnetic separation

Magnetic separation, often defined as magnetophoresis, is widely exploited in LOC applications. Magnetophoresis is a nondestructive method for the selective collection or separation of magnetic particles, by moving them in a viscous medium under the influence of an applied magnetic field [16]. Usually, in LOC applications, we refer to free-flow magnetophoresis, since the separation of particles or magnetic objects takes place in a liquid environment where magnetic particles are deflected from the direction of laminar flow by a perpendicular magnetic field (**Figure 3**); the extent of the deflection depends mainly on flow rate and on the susceptibility of the magnetic particle, or more precisely, on the susceptibility mismatch between the particle and the fluid.

The vector u_{defl}, which represents the deflection of magnetic particles due to the applied magnetic field, is the result of two contributions: the flow velocity induced on the particle by the applied magnetic field u_{mag}, and the hydrodynamic flow velocity u_{hyd}:

$$u_{defl} = u_{mag} + u_{hyd} \tag{9}$$

The magnetically induced flow velocity u_{mag} can be expressed as the ratio of the magnetic force F_m exerted on the particle to the viscous drag force:

$$u_{mag} = \frac{F_m}{F_d} = \frac{F_m}{6\pi\eta r} \tag{10}$$

In a magnetophoresis application, the magnetic force depends on the particle features, mainly its volume V_p, on the mismatch in terms of magnetic properties between the particle and the fluid, and on the applied magnetic field B. Eq. (5) becomes consequently [17]:

$$F_m = \frac{V_p \cdot \Delta\chi}{\mu_0}(B \cdot \nabla)B \tag{11}$$

Eq. (11) is suitable for both paramagnetic and superparamagnetic particles, since soft magnetism approximation, and lack of magnetic memory is considered for the particles, and for relatively high magnetic field strengths, able to induce in the particles a magnetization close to the saturation one. To this aim, macroscopic permanent magnets and electromagnets can be exploited, since they produce sufficiently strong fields (>0.5 T), able to saturate superparamagnetic particles.

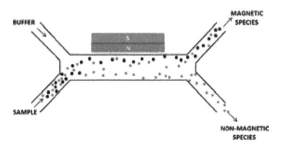

Figure 3. Typical schematization of a magnetophoresis scenario in which a magnetic field perpendicular to the flow direction is exploited to deviate magnetic particles from the trajectory imposed by fluidic forces and thus to separate them from the rest of the sample. Complete separation of the species can be obtained through some consecutive steps.

It is possible to distinguish between positive and negative magnetophoresis depending on the sign of the susceptibility mismatch: if $\Delta\chi$ is positive, for example, in the case of magnetic micro-/

nano-objects in a non-magnetic fluid, we talk about positive magnetophoresis and the particles are attracted by the externally applied magnetic field. On the other hand, when the suscepti-bility mismatch $\Delta\chi$ is negative, for example, in the case of diamagnetic particles in a magne-tofluid, the particles are repelled from the magnetic field and negative magnetophoresis occurs. When designing a magnetophoresis device, it is necessary to assure the dominance of magnetic forces respect to the other physical contributions. Considering Eq. (11), F_m results proportional to the magnetic field gradients and to the susceptibility mismatch. Several strategies aiming at optimizing magnetophoresis have been investigated and proposed, and they may be essentially classified in two categories distinguishing between those aiming at maximizing the magnetic field gradients and those acting on the susceptibility mismatch [18].

In the attempt to increase the magnetic field gradient, many solutions have been proposed in literature, all aiming at the generation of a nonuniform magnetic field distribution. In some cases, uniform external magnetic field sources, such as permanent magnets, were combined with wires [19], ferromagnetic strings [20], or magnetic microparticles embedded in the chip structure itself [21], whereas in other cases, the integration of small electromagnets produced intense magnetic field gradients [22]. Alternatively, permanent magnets can be actuated to generate a time-dependent magnetic field, for example, through the use of a rotating magnetic wheel [23], or arranged in an asymmetric configuration, thus generating spatial field variation or multiple magnetic separation stages [24]. One of the most commonly employed magnet configuration in magnetophoresis applications, able to maximize field distribution anisotropy, is the quadrupolar arrangement which creates a magnetic gradient radially outward from the center of the flow column [25]. In a variant of this, the fluid is static, while an applied magnetic field is moved up the container [26]. The particles move in the resulting field gradient at a velocity dependent on their magnetophoretic mobility. At the top of the container, they enter a removable section and are held here by a permanent magnet. The bottom section of the container moves to the next section where a magnetic field with different strength to the first is applied and the process repeats. The result is a fractionation of the sample into aliquots differing for magnetophoretic mobility [27].

The alternative strategy to enhance the magnetic separation capabilities of the device lies in increasing the susceptibility mismatch by modifying either the susceptibility of the particle, or the one of the surrounding fluid. This can be accomplished (1) by labeling the cells or the desired microstructure with higher magnetic susceptibility beads (2) by internalizing higher quantity of magnetic material, and (3) by using a ferrofluid medium, for example, by adding gadolinium, and diamagnetic particles instead of the para/ferromagnetic ones [28].

4.2. Magnetic micromanipulation

Magnetic fields can be employed in LOC for the non-contact manipulation of biological samples or other magnetically labeled substances/structures. Apart from applications in which it is necessary to separate different types of samples or specific entities from the medium (tasks that can benefit from techniques mainly based on magnetophoresis), in some cases, precise manipulation or transport along defined paths is required. This kind of task is more complex in terms of extent of magnetic fields required and of control, considering that for 3D manipulation, torque equilibrium must be taken into account, in addition to force balance.

It has been demonstrated that biological systems or chemical samples labeled with magnetic nanoparticles can be micro-/nano-manipulated or transported in three dimensions, by exploiting combinations of electromagnets or permanent magnets, possibly moved or rotated along three axes. Permanent magnets offer the advantage to produce large fields without the need of any electrical current, thus avoiding powering, heating and control issues, which have to be faced instead when using electromagnets. On the other hand, electromagnets offer the possibility to tune magnetic field gradients and field intensities by simply varying the currents across the coils. By properly combining electromagnets, it is possible to produce in the workspace both varying magnetic fields, without the need of moving parts, and spatially uniform magnetic fields and gradients. This makes possible to accomplish also quite complex manipulation and locomotion paradigms. Several architectures have been proposed, presenting different magnet and electromagnet arrangements. In LOC applications, due to the need to finely control the locomotion of small-scale entities, electromagnets are the most commonly employed solution.

Generally, when an electrical current flows in a wire, a magnetic field is generated according to the Biot–Savart theory [9]. When considering a single circular coil in which a current with magniture I is flowing, the magnetic field along the central axis of the coil can be defined as:

$$B = \quad dB = \frac{Ir^2}{2(r^2 + z^2)^{\frac{3}{2}}} \qquad (12)$$

where r is the radius of the coil, and z is the coordinate along the central axis.

Often, specific coil pairs arrangements, namely Helmholtz and Maxwell coils, are exploited in micromanipulation applications. They consist of two identical circular magnetic coils symmetrically placed along a common axis, one on each side of the workspace, and separated by a distance d corresponding to coil radius (r_H) in the case of Helmhotz coils and to $\sqrt{3}r_M$ in the case of Maxwell coils. In Helmholtz arrangement, each coil carries an equal electric current in the same direction, whereas in Maxwell coils currents flow in opposite directions. The magnetic flux density in case of Helmholtz and Maxwell coils can be derived from Eq. (12):

$$B_H = \frac{\mu_0 r_H^2 N_H I_H}{2} \left(\frac{1}{\left[r^2 + \left(\frac{d}{2} - z \right)^2 \right]^{\frac{3}{2}}} + \frac{1}{\left[r^2 + \left(\frac{d}{2} + z \right)^2 \right]^{\frac{3}{2}}} \right) \qquad (13)$$

$$B_M = \frac{\mu_0 r_M^2 N_M I_M}{2} \left(\frac{1}{\left[r^2 + \left(\frac{d}{2} - z \right)^2 \right]^{\frac{3}{2}}} - \frac{1}{\left[r^2 + \left(\frac{d}{2} + z \right)^2 \right]^{\frac{3}{2}}} \right) \tag{14}$$

N_H, I_H, r_H, N_M, I_M, r_M are the numbers of windings, current, and radius of Helmholtz and Maxwell coils, respectively.

When considering a combination of Helmholtz and Maxwell coils (**Figure 4A**), the magnetic field B and magnetic field gradient ∇B in the workspace can be derived analytically by Eqs. (13) and (14) as follows:

$$B = \frac{8\mu_0 N_H I_H}{5\sqrt{5} r_H} \tag{15}$$

$$B = \frac{48\sqrt{3} \mu_0 N_M I_M}{49\sqrt{7} r_M^2} \tag{16}$$

Equations (15) and (16) clearly show that Helmholtz coils are able to generate a uniform magnetic field, whereas Maxwell coils produce a uniform magnetic field gradient along its axis. For this reason, combinations of Helmholtz and Maxwell coils have been exploited to obtain both a uniform field gradient and magnetic field uniformity [29].

Nonuniform field setups have been developed as well. Despite the major complexity both in terms of design/fabrication and control, they enable an increase in the number of controlled degrees of freedom. In this sense, a representative example is the OctoMag system [30, 31] (**Figure 4B**), designed for the control of intraocular microrobots for minimally invasive retinal therapy and diagnosis, but showing also potentialities for use as a wireless micromanipulation apparatus. It consists of eight stationary electromagnets with soft-magnetic cores able to generate predefined values of magnetic field and gradient, providing the manipulated object with five degrees of freedom; this system can operate closed-loop position control by exploiting computer-assisted visual tracking or in open loop by relying only on the operator microscope-mediated visual feedback. Alternative approaches aim at exploiting other sources of nonuniform magnetic fields: Martel et al. [32] demonstrated the effectiveness of using an MRI scanner for the control of a swarm of magnetotactic bacteria in executing a manipulation task on micro-objects. Micro-assembly of micro-objects using a cluster of microparticles (with average diameter of 100 μm) and a magnetic-based manipulation system has also been shown in [33].

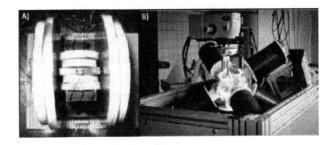

Figure 4. Different magnetic field generation setup exploited for magnetic manipulation. (A) Combination of Helmholtz and Maxwell coils; (B) OctoMag system [31].

5. Applications of magnetic fields in lab-on-a-chip

Once provided the reader with an overview of physics at the microscale, as well as of magnetic materials properties and principles to be exploited, some real applications of LOCs will be described, where using magnetic fields. Such applications range from biological samples handling to chemical reactions and other manipulation tasks. As a consequence, in order to better analyze the potential of using magnetic fields in this context, it is useful to classify the applications in three main areas: (1) on-chip bioanalysis; (2) cell separation or manipulation; and (3) non-conventional manipulation techniques.

5.1. On-chip bioanalysis

A vast number of reactions in genomics, proteomics, and clinical medicine need molecular mixing of fluids or recognition events between single strands of DNA, between antibodies and antigens, or between receptors and cells. Such reactions usually require a number of steps that must be performed sequentially, such as isolation, washing, or purification. In this kind of applications, the introduction of automation could lead to higher throughput and the use of magnetic fields, commonly mediated by the use of magnetic particles, has revealed to be extremely useful in some on-chip functions such as the mixing of fluids, selective capture of analytes, later to be transferred for further analysis steps, or the performance of stringency and washing. Usually two main properties of magnetic particles are exploited in analytical assays: the possibility to biofunctionalize them, thus enabling selective binding and related applications, and the capability to form supra-particle structures, such as chains, exploited mainly in fluid mixing and analytes capture applications [5].

Usually, magnetic particles are labeled with molecules, for example, antibodies, showing high affinity for the target species and able to mediate the binding with them. Often multistep binding processes are carried out to bind the particle–analyte complex, for example to fluorescent dies, thus to enable target detection [34]. In other cases, for additional purification steps after labeling, magnetic separation, mediated by magnetophoresis phenomena, or transfer processes are required to enable further purification/washing steps or analysis. This kind of procedures can be exploited both for the in vitro purification of nucleic acids or proteins and for biomolecules separation, as well as DNA sequencing. For example, in case of DNA extraction and separation processes, magnetic particles are firstly held in place by exploiting external magnetic fields, thus exposing their functional groups to specific DNA strands. Magnetic separation steps can consequently be performed to isolate the strands of interest from the rest of the sample [35]. In other cases, the target-binding capabilities, with analytes showing at least two epitopes, have been exploited to create large aggregates for albumin detection in buffer [36]: The binding between particles is mediated by the target molecule and thus the extent of the aggregate represents a measure of analytes concentration within the solution and the magnetic properties of the aggregate allow their detection. In this case, non-specific particle clustering should be avoided.

Another interesting application is in the field of biosensing or surface binding bioassays: In this context, magnetic particles can be exploited as binding mediators between the target species and a functionalized surface. The most common configuration in this field is the sandwich one, in which the target molecule binds first the magnetic particles dispersed in the solution. Later, by exploiting magnetic field gradients, the complex can be moved toward the surface of the sensor where interactions at the molecular scale occur [37].

Whereas in detection applications, non-specific magnetic interaction within particles has to be avoided, there are applications in which the capability of magnetic particles to interact each other thanks to magnetic forces to form supra-particle structures can be advantageously exploited. When working with really precious/expensive fluids, really small volumes and microfluidic devices are employed. Due to the small scale and to the strong viscous force, fluid mixing is not straightforward. In this case, magnetic particles can be exploited to steer fluids. This is done specifically for supra-particle structures such as chains created by the dipole–dipole interactions between magnetic particles, and actuated, for example, by means of rotating magnetic fields [38].

5.2. Magnetic cell separation

Cell manipulation by means of magnetic fields relies on cells magnetic labeling or, alternatively, on magnetic particles internalization through magnetic field-mediated transfection mechanisms (magnetofection). In a typical in vitro magnetofection system, target cells are located at the bottom of a fluidic chamber well of a culture plate, and a permanent magnet beneath the chamber provides a magnetic force that attracts the biofunctional particles toward the cells.

Separation and isolation of rare cell populations from a heterogeneous suspension is essential for many applications, ranging from disease diagnostics to drug screening. Various separation techniques have been proposed, but magnetic fields emerged as very promising also in this kind of application thanks to the exploitation of the magnetic separation principles presented in the previous section.

Magnetic cell sorting can be operated in either a serial or a parallel manner, resulting in higher throughput with up to 10^{11} cells processed in 30 min. This process can be operated in both batch and continuous flow mode. In batch processing, the hardware is very simple, including a magnetic field source placed close to a column containing the cells to be separated. Several architectures were developed to this aim both at large scales, exploiting for example ferromagnetic columns [39] and at smaller scales with arrays of electrical wires exploited to produce local magnetic fields [40]. In the case of continuous flow cell sorting, instead, typical magnetophoresis principles are exploited.

Multicell sorting systems rely on the variation in the uptake of magnetic material between different cell populations and thus on the different path deviation produced by magnetic field gradients. They can be also used for the separation of different cell species from heterogeneous samples [41].

The separation of a specific class of cells from a certain sample is extremely important for some applications, for example, for the detection of pathology or for the testing of a therapeutic strategy. For example, diagnosis and treatment of HIV disease rely on the efficient separation of human T-lymphocytes from whole blood [42], whereas in the diagnosis and treatment of malaria, the detection of infected red blood cells (RBCs) and their separation from healthy cells is mandatory [43]. Separation of neuronal cells has gained interest for its potential applications in cell replacement therapy of neurodegenerative disorders such as Parkinson's disease, multiple sclerosis, and Alzheimer's disease [44]. Cell separation methods are also needed for separating nucleated RBCs from the peripheral blood of pregnant women, for monitoring maternal, fetal, and neonatal health [45].

Magnetic field-based cell counting techniques have also been developed. One method estimates the location and number of cells tagged by measuring the magnetic moment of the microsphere tags [46], while another uses a giant magnetoresistive sensor to measure the location of microspheres attached to a surface layered with a bound analyte [47].

5.3. Non-conventional manipulation strategies based on magnetic fields

When high sensitivity, not compatible with magnetophoretic techniques is required, and independence on the human operator are desirable, microrobotic manipulators acting at the cellular scale can offer significant benefits. Wirelessly controlled (i.e., untethered) cell-sized robots are highly noninvasive. At this length scale, where viscous fluid forces dominate inertial ones, mobile microrobots cause very little mixing or agitation of the surrounding environment. This is a significant advantage, for example, over suction pipetting for life scientists, since pipettes cause relatively large fluid disturbances [48]. Magnetic control of microrobots and microgrippers is gaining growing importance in micro-object manipulation: in addition to increasing the manipulation accuracy, the exploitation of such micro-systems avoids sometimes the direct magnetization of the sample, through internalization or labeling, thus helping in keeping its integrity. Many challenges have to be faced to enable single cell manipulation. When working with single cells or with really fragile samples, in fact, it is essential to have microstructures with sizes comparable to those ones of the target, to be able to finely control them within the workspace, and to avoid to affect cell viability or samples integrity due to the microrobot exploitation.

Some research groups focusing on microtechnologies have been working toward a high efficiency in vitro fertilization (IVF) process [49] (**Figure 5A**). The IVF goal is to fertilize oocytes, and it consists of several manually or teleoperated manipulation steps that require important practical skills. Sakar et al. [50] developed microtransporters using a simple, single-step microfabrication technique allowing parallel fabrication. They demonstrated that the micro-transporters can be navigated to separate individual targeted cells with micron-scale precision and deliver microgels without disturbing the cells in the neighborhood and the local micro-environment. Yamanishi et al. [51] presented an innovative driving method, devised for cell sorting, for an on-chip robot actuated by permanent magnets in a chip, where a piezoelectric ceramic is applied to induce ultrasonic vibration to the microfluidic chip and the high-frequency vibration reduces significantly the effective friction on a magnetically driven microtool.

Other interesting magnetic microstructures, devised for cell manipulation in in vitro environments for LOC applications, but finally eligible in the future for in vivo applications, have been recently proposed. Examples are novel microgrippers, in which both the navigation and the gripper actuation rely on magnetic fields [52] (**Figure 5B**), 3D laser lithography microcages devised to act as cell carriers (**Figure 5D**) [53] or thin magnetic films working at the air fluid

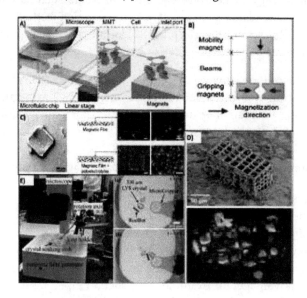

Figure 5. Overview of non-conventional manipulation systems for microrobotics and LOC applications. (A) Conceptual overview of a microfluidic cell manipulation system based on magnetically driven microtools and exploited for oocytes handling [49] (reproduced with permission from Royal Society of Chemistry); (B) Schematic representation of a remotely controlled microgripper exploiting magnetic fields both for navigation and for gripper actuation (adapted from [52] and reproduced with permission from Royal Society of Chemistry); (C) SEM image of a magnetic thin film devised for cell manipulation (left) and schematic representation of the film structure with microscope images showing T24 cell compatibility with the magnetic structure (right) [13]; (D) SEM image (above) and confocal microscope image of a magnetic microcage after cell culture [53]; (E) Experimental setup for magnetic micromanipulation (left) and microscope images of the magnetic microrobot during christal manipula-tion tasks (adapted from [55] and reproduced with permission of the International Union of Crystallography).

interface and exploiting surface tension phenomena together with magnetic navigation and showing compatibility with cell manipulation applications [54] (**Figure 5C**). Interesting is also the development of manipulation strategies for precise non-contact handling of small and fragile samples based on complex control algorithms aiming at creating vortexes, as demonstrated for crystal harvesting applications (**Figure 5E**) [55].

6. Conclusions

LOC technologies represented a real breakthrough in the last decades for in vitro laboratory analyses. However, the integration of really complex functions in a limited space results extremely challenging and further efforts are required to make LOC systems accurate and operating in an automated fashion. Magnetic fields exploitation revealed to be extremely promising and effective in the execution of certain tasks, with the aim of overcoming some of the limitations connected to human operators and enabling procedures impossible with traditional laboratory techniques.

In this sense, the role played by magnetic nanoparticles is extremely important, but alternative techniques providing the samples to be manipulated with magnetic properties have been investigated and show great potentialities.

In some cases, magnetic field-based technologies appear more advantageous compared with other LOC actuation strategies, first of all the fluidic one. However, in view of more reliable systems, a possible future trend, already investigated in many applications focuses on combining several effects, including chemical binding, microfluidic actuation, magnetic and electric fields, to obtain more efficient analytical and biological testing platforms. A further enhancement of LOC devices, and especially of those exploiting magnetic fields, may derive from the integration of technologies that are typical of the microrobotics world. Some examples have been reported in the Section 5.3 and an interesting contribution could derive from microrobotics, both in terms of cell carriers and manipulation systems fabrication, and in terms of control strategies.

The development of cheaper and more reliable LOCs could enable many steps forward in really important fields, such as nanomedicine, personalized medicine, and cellular studies. The advantages and technological progresses offered by magnetic technologies at all the scales and in different fields could surely help to reach this goal.

Author details

Veronica Iacovacci*, Gioia Lucarini, Leonardo Ricotti and Arianna Menciassi

*Address all correspondence to: v.iacovacci@sssup.it

The BioRobotics Institute, Scuola Superiore Sant'Anna, Pisa, Italy

References

[1] Neuži P, Giselbrecht S, Länge K, Huang TJ, Manz A. Revisiting lab-on-a-chip technology for drug discovery. Nature Reviews Drug Discovery. 2012;11(8):620–32. doi: 10.1038/nrd3799

[2] Abgrall P, Gue AM. Lab-on-chip technologies: making a microfluidic network and coupling it into a complete microsystem—a review. Journal of Micromechanics and Microengineering. 2007;17(5):R15. doi:10.1088/0960-1317/17/5/R01.

[3] Bhagat AA, Bow H, Hou HW, Tan SJ, Han J, Lim CT. Microfluidics for cell separation. Medical & Biological Engineering & Computing. 2010;48(10):999–1014. doi:10.1007/s11517-010-0611-4

[4] Abbott J, Nagy Z, Beyeler F, Nelson B. Robotics in the small. IEEE Robotics and Automation Magazine. 2007 J;14:92–103. doi:10.1109/MRA.2007.380641

[5] van Reenen A, de Jong AM, den Toonder JM, Prins MW. Integrated lab-on-chip biosensing systems based on magnetic particle actuation—a comprehensive review. Lab on a Chip. 2014;14(12):1966–86. doi:10.1039/C3LC51454D

[6] Furlani EP, Ng KC. Analytical model of magnetic nanoparticle transport and capture in the microvasculature. Physical Review E. 2006;73(6):061919. doi:10.1103/PhysRevE.73.061919

[7] Jiles D. Introduction to magnetism and magnetic materials. Taylor and Francis Group CRC Press; 2015.

[8] Martel S. Magnetic nanoparticles in medical nanorobotics. Journal of Nanoparticle Research. 2015;17(2):1–5. doi:10.1007/s11051-014-2734-2

[9] Furlani EP. Permanent magnet and electromechanical devices: materials, analysis, and applications. Academic Press; 2001.

[10] Zborowski M, Ostera GR, Moore LR, Milliron S, Chalmers JJ, Schechter AN. Red blood cell magnetophoresis. Biophysical Journal. 2003;84(4):2638–45. doi:10.1016/S0006-3495(03)75069-3

[11] Plouffe BD, Murthy SK, Lewis LH. Fundamentals and application of magnetic particles in cell isolation and enrichment: a review. Reports on Progress in Physics. 2014;78(1):016601. doi:10.1088/0034-4885/78/1/016601

[12] Lewin M, Carlesso N, Tung CH, Tang XW, Cory D, Scadden DT, Weissleder R. Tat peptide-derivatized magnetic nanoparticles allow in vivo tracking and recovery of progenitor cells. Nature Biotechnology. 2000;18(4):410–4. doi:10.1038/74464

[13] Iacovacci V, Lucarini G, Innocenti C, Comisso N, Dario P, Ricotti L, Menciassi A. Polydimethylsiloxane films doped with NdFeB powder: magnetic characterization and

potential applications in biomedical engineering and microrobotics. Biomedical Microdevices. 2015;17(6):1–7. doi:10.1007/s10544-015-0024-0

[14] Canham L. Handbook of porous silicon. Springer; 2014.

[15] Spinu, L, Dodrill BC, Radu C. Magnetometry measurements. Magnetics Technology International. 2013 pp. 62–65

[16] Yan H, Wu H. Magnetophoresis. Encyclopedia of Microfluidics and Nanofluidics. 2015:1696–701.

[17] Shevkoplyas SS, Siegel AC, Westervelt RM, Prentiss MG, Whitesides GM. The force acting on a superparamagnetic bead due to an applied magnetic field. Lab on a Chip. 2007;7(10):1294–302. doi:10.1039/B705045C

[18] Hejazian M, Li W, Nguyen NT. Lab on a chip for continuous-flow magnetic cell separation. Lab on a Chip. 2015;15(4):959–70. doi:10.1039/C4LC01422G

[19] Jung J, Han KH. Lateral-driven continuous magnetophoretic separation of blood cells. Applied Physics Letters. 2008;93(22):223902. doi:10.1063/1.3036898

[20] Adams JD, Kim U, Soh HT. Multitarget magnetic activated cell sorter. Proceedings of the National Academy of Sciences. 2008;105(47):18165–70. doi:10.1073/pnas.0809795105

[21] Faivre M, Gelszinnis R, Degouttes J, Terrier N, Riviere C, Ferrigno R, Deman AL. Magnetophoretic manipulation in microsystem using carbonyl iron-polydimethylsiloxane microstructures. Biomicrofluidics. 2014;8(5):054103. doi:10.1063/1.4894497

[22] Liu C, Lagae L, Borghs G. Manipulation of magnetic particles on chip by magnetophoretic actuation and dielectrophoretic levitation. Applied Physics Letters. 2007;90(18):184109. doi:10.1063/1.2736278

[23] Verbarg J, Kamgar-Parsi K, Shields AR, Howell PB, Ligler FS. Spinning magnetic trap for automated microfluidic assay systems. Lab on a Chip. 2012;12(10):1793–9. doi:10.1039/C2LC21189K

[24] Jung Y, Choi Y, Han KH, Frazier AB. Six-stage cascade paramagnetic mode magnetophoretic separation system for human blood samples. Biomedical Microdevices. 2010;12(4):637–45. doi:10.1007/s10544-010-9416-3

[25] Moore LR, Rodriguez AR, Williams PS, McCloskey K, Bolwell BJ, Nakamura M, Chalmers JJ, Zborowski M. Progenitor cell isolation with a high-capacity quadrupole magnetic flow sorter. Journal of Magnetism and Magnetic Materials. 2001;225(1):277–84. doi:10.1016/S0304-8853(00)01251-8

[26] Todd P, Cooper RP, Doyle JF, Dunn S, Vellinger J, Deuser MS. Multistage magnetic particle separator. Journal of Magnetism and Magnetic Materials. 2001;225(1):294–300. doi:10.1016/S0304-8853(00)01253-1

[27] Pankhurst QA, Connolly J, Jones SK, Dobson JJ. Applications of magnetic nanoparticles in biomedicine. Journal of Physics D: Applied Physics. 2003;36(13):R167. doi: 10.1088/0022-3727/36/13/201

[28] Zeng J, Deng Y, Vedantam P, Tzeng TR, Xuan X. Magnetic separation of particles and cells in ferrofluid flow through a straight microchannel using two offset magnets. Journal of Magnetism and Magnetic Materials. 2013;346:118–23. doi:10.1016/j.jmmm. 2013.07.021

[29] Jeong S, Choi H, Choi J, Yu C, Park JO, Park S. Novel electromagnetic actuation (EMA) method for 3-dimensional locomotion of intravascular microrobot. Sensors and Actuators A: Physical. 2010;157(1):118–25. doi:10.1016/j.sna.2009.11.011

[30] Kummer MP, Abbott JJ, Kratochvil BE, Borer R, Sengul A, Nelson BJ. OctoMag: An electromagnetic system for 5-DOF wireless micromanipulation. IEEE Transactions on Robotics. 2010;26(6):1006–17. doi:10.1109/TRO.2010.2073030

[31] Ullrich F, Bergeles C, Pokki J, Ergeneman O, Erni S, Chatzipirpiridis G, Pané S, Framme C, Nelson BJ. Mobility experiments with microrobots for minimally invasive intraocular Surgery Microrobot experiments for intraocular surgery. Investigative Ophthalmology & Visual Science. 2013;54(4):2853–63. doi:10.1167/iovs.13-11825

[32] Martel S, Mohammadi M, Felfoul O, Lu Z, Pouponneau P. Flagellated magnetotactic bacteria as controlled MRI-trackable propulsion and steering systems for medical nanorobots operating in the human microvasculature. The International Journal of Robotics Research. 2009;28(4):571–82. doi:10.1177/0278364908100924

[33] Khalil IS, Magdanz V, Sanchez S, Schmidt OG, Misra S. Three-dimensional closed-loop control of self-propelled microjets. Applied Physics Letters. 2013;103(17):172404. doi: 10.1063/1.4826141

[34] Lacharme F, Vandevyver C, Gijs MA. Full on-chip nanoliter immunoassay by geometrical magnetic trapping of nanoparticle chains. Analytical Chemistry. 2008;80(8):2905–10. doi:10.1021/ac7020739

[35] Liu P, Li X, Greenspoon SA, Scherer JR, Mathies RA. Integrated DNA purification, PCR, sample cleanup, and capillary electrophoresis microchip for forensic human identification. Lab on a Chip. 2011;11(6):1041–8. doi:10.1039/C0LC00533A

[36] Moser Y, Lehnert T, Gijs MA. On-chip immuno-agglutination assay with analyte capture by dynamic manipulation of superparamagnetic beads. Lab on a Chip. 2009;9(22):3261–7. doi:10.1039/B907724C

[37] Dittmer WU, De Kievit P, Prins MW, Vissers JL, Mersch ME, Martens MF. Sensitive and rapid immunoassay for parathyroid hormone using magnetic particle labels and magnetic actuation. Journal of Immunological Methods. 2008;338(1):40–6. doi:10.1016/j.jim.2008.07.001

[38] Gao Y, Beerens J, van Reenen A, Hulsen MA, de Jong AM, Prins MW, den Toonder JM. Strong vortical flows generated by the collective motion of magnetic particle chains rotating in a fluid cell. Lab on a Chip. 2015;15(1):351–60. doi:10.1039/C4LC01198H

[39] Miltenyi S, Müller W, Weichel W, Radbruch A. High gradient magnetic cell separation with MACS. Cytometry. 1990;11(2):231–8. doi:10.1002/cyto.990110203

[40] Lee H, Purdon AM, Westervelt RM. Manipulation of biological cells using a micro-electromagnet matrix. Applied Physics Letters. 2004;85:1063. doi:10.1063/1.1776339

[41] Pamme N, Wilhelm C. Continuous sorting of magnetic cells via on-chip free-flow magnetophoresis. Lab on a Chip. 2006;6(8):974–80. doi:10.1039/B604542A

[42] Cheng X, Irimia D, Dixon M, Sekine K, Demirci U, Zamir L, Tompkins RG, Rodriguez W, Toner M. A microfluidic device for practical label-free CD4+ T cell counting of HIV-infected subjects. Lab on a Chip. 2007;7(2):170–8. doi:10.1039/B612966H

[43] Paul F, Melville D, Roath S, Warhurst DC. A bench top magnetic separator for malarial parasite concentration. IEEE Transactions on Magnetics. 1981;17(6):2822–4. doi:10.1109/TMAG.1981.1061711

[44] Wu Z, Hjort K, Wicher G, Svenningsen ÅF. Microfluidic high viability neural cell separation using viscoelastically tuned hydrodynamic spreading. Biomedical Microdevices. 2008;10(5):631–8. doi:10.1007/s10544-008-9174-7

[45] Huang R, Barber TA, Schmidt MA, Tompkins RG, Toner M, Bianchi DW, Kapur R, Flejter WL. A microfluidics approach for the isolation of nucleated red blood cells (NRBCs) from the peripheral blood of pregnant women. Prenatal Diagnosis. 2008;28(10):892–9. doi:10.1002/pd.2079

[46] Hofmann WK, de Vos S, Komor M, Hoelzer D, Wachsman W, Koeffler HP. Characterization of gene expression of CD34+ cells from normal and myelodysplastic bone marrow. Blood. 2002;100(10):3553–60. doi:10.1182/blood.V100.10.3553

[47] Edelstein RL, Tamanaha CR, Sheehan PE, Miller MM, Baselt DR, Whitman L, Colton RJ. The BARC biosensor applied to the detection of biological warfare agents. Biosensors and Bioelectronics. 2000;14(10):805–13. doi:10.1016/S0956-5663(99)00054-8

[48] Brehm-Stecher BF, Johnson EA. Single-cell microbiology: tools, technologies, and applications. Microbiology and Molecular Biology Reviews. 2004;68(3):538–59. doi:10.1128/MMBR.68.3.538-559.2004

[49] Hagiwara M, Kawahara T, Yamanishi Y, Masuda T, Feng L, Arai F. On-chip magnetically actuated robot with ultrasonic vibration for single cell manipulations. Lab on a Chip. 2011;11(12):2049–54. doi:10.1039/C1LC20164F

[50] Sakar MS, Steager EB, Cowley A, Kumar V, Pappas GJ. Wireless manipulation of single cells using magnetic microtransporters. In: 2011 IEEE International Conference on

Robotics and Automation (ICRA), (pp. 2668–2673). IEEE. doi:10.1109/ICRA. 2011.5980100

[51] Yamanishi Y, Sakuma S, Onda K, Arai F. Powerful actuation of magnetized microtools by focused magnetic field for particle sorting in a chip. Biomedical Microdevices. 2010;12(4):745–52. doi:10.1007/s10544-010-9428-z

[52] Chung SE, Dong X, Sitti M. Three-dimensional heterogeneous assembly of coded microgels using an untethered mobile microgripper. Lab on a Chip. 2015;15(7):1667–76. doi:10.1039/C5LC00009B

[53] Kim S, Qiu F, Kim S, Ghanbari A, Moon C, Zhang L, Nelson BJ, Choi H. Fabrication and characterization of magnetic microrobots for three-dimensional cell culture and targeted transportation. Advanced Materials. 2013;25(41):5863–8. doi:10.1002/adma. 201301484

[54] Lucarini G, Iacovacci V, Ricotti L, Comisso N, Dario P, Menciassi A. Magnetically driven microrobotic system for cancer cell manipulation. In Engineering in Medicine and Biology Society (EMBC), 2015 37th Annual International Conference of the IEEE 2015 (pp. 3631–3634). IEEE. doi:10.1109/EMBC.2015.7319179

[55] Tung HW, Sargent DF, Nelson BJ. Protein crystal harvesting using the RodBot: a wireless mobile microrobot. Journal of Applied Crystallography. 2014;47(2):692–700. doi:10.1107/S1600576714004403

2

Microfluidic Multiple Chamber Chip Reactor Filled with Enzyme-Coated Magnetic Nanoparticles

Ferenc Ender, Diána Weiser and László Poppe

Abstract

In this chapter, a novel microfluidic device (MagneChip) is described which comprises microliter volume reaction chambers filled with magnetically fixed enzyme-coated magnetic nanoparticles (ecMNPs) and with an in-line UV detector. In the experiments, MNPs with phenylalanine ammonia-lyase (PAL)—an enzyme which catalyzes the deamination of l-phenylalanine (Phe) to (E)-cinnamate in many organisms—immobilized on the surface were applied as biocatalyst to study the characteristics of the MagneChip device. In the reaction chambers of this microfluidic device, the accurate in situ quantization of the entrapped MNPs was possible using a resonant coil magnetometer integrated below the chambers. Computational fluid dynamics (CFD) calculations were used to simulate the flow field in the chambers. The enzyme-catalyzed biotransformations could be performed in the chip with excellent reproducibility and of repeatability. The platform enabled fully automatic multiparameter measurements with a single biocatalyst loading of about 1 mg PAL-ecMNP in the chip. A study on the effect of particle size and arrangement on the catalytic activity revealed that the mass of ecMNPs fixed in the chamber is independent of the particle diameter. Decreasing the particle size resulted in increasing catalytic activity due to the increased area to volume ratio. A binary mixture of particles with two different particle sizes could increase the entrapped particle mass and further the catalytic activity compared to the best uniform packing. The platform enabled a study of biotransformation of l-phenylalanine and five unnatural substrates by consecutive reactions using same PAL-ecMNP loading. With the aid of the platform, we first demonstrated that PAL can catalyze the ammonia elimination from the noncyclic propargylglycine as substrate.

Keywords: Magnetic nanoparticles, Magnetic chip reactor, Microfluidic reactor, Enzyme reaction, Phenylalanine ammonia-lyase

1. Introduction

Microreactors are usually defined as miniaturized reaction systems fabricated using methods of microtechnology and precision engineering. The term "microreactor" is the proposed name for a wide range of devices, having typically submillimeter channel dimensions which can be further divided into submicron sized components, for example, microparticle and nanoparticle carriers [1].

Before evolution of microreactor technology, the traditional way to conduct solution phase synthesis and analysis was the batch mode in stationary reactors with stirring or shaking to mix the reactants. Nowadays, microstructured devices offer greatly enhanced performance compared with conventional batch systems due to effects arising from the microscale domain:

- Batch processes are space-resolved; therefore, the process must be readjusted in each demand for larger product quantities. In contrast, flow microreactor processes are time-resolved; therefore, the output of the reaction is determined by the flow rate and the operation time, and no further optimization is needed. This also leads to accelerated process development and enhanced safety due to smaller reactor volumes [2].

- Microreactors with high surface-to-volume ratio (SVR) are able to absorb the heat evolving in an exothermic reaction more efficiently than any batch reactor. Therefore, the temperature distribution inside the microreactor is homogenous in the whole volume. In contrast, small SVR usually leads to uneven temperature distribution in large-scale batch reactors, decreasing the product yield [2].

- Mixing quality is crucial for many reactions, where the molar ratio between the reactants needs to be controlled precisely. Short diffusion paths provide efficient mixing in microreactors, which overrides the achievable mixing efficiency of batch reactors [2].

- In biocatalytic applications, the efficiency of the microreactor can be further improved by immobilization of enzymes on nanoscale carriers accommodating in the reactor. Reusability of the biocatalyst makes the process economical and more environmentally friendly.

- To perform similar analyses in shorter timescale even in parallel is an anticipated objective for screening and routine use in protein and enzyme research [3]. A desirable goal is the high throughput screening of enzymes and their substrates and inhibitors. The prospective fields of application of microreactors are quite wide and include biotechnology, as well as combinatorial chemistry and enzyme-targeted drug discovery [4].

- Analytical systems which comprise microreactors are characterized by outstanding repeatability and reproducibility, due to replacing iterative steps in batch and discrete sample treatment by flow injection systems [4]. Benefitting from system automation, this also eliminates errors associated with manual protocols.

- Small reagent volume is also a benefit of microreactors enabling economical and efficient screening of novel reaction paths and substrates.

- Microreactors have a high potential in industry, as developments by microreactors can be faster transferred into production at lower costs than batch processes.

Despite of the rapid development of enzymatic microreactors in the recent decade, important design questions still need to be answered.

Reaction kinetics is a key parameter of device design. Widely used kinetic parameters are deduced from the Michaelis–Menten model, which is valid only in batch reactions. In flow systems, the flow effects should be also considered. Immobilization of the enzymes causes further complication in modeling. Immobilization may affect the intrinsic kinetic parameters and may influence the availability of the enzyme. The kinetic model should also consider that the liquid phase containing the substrate and product is moving compared to the solid phase containing the immobilized enzyme.

When supported catalyst-filled microreactors are used, reproducible filling of the supported catalyst into the reactor space is not always straightforward. Even more challenging is the quantification of the actual load of the carriers.

Long-term stability of the reactor and the reproducibility of the measurements may be affected —among other factors—by the flow rate, the substrate concentration, and the morphology of the immobilized biocatalyst.

This chapter presents results carried out by a microfluidic microreactor system, the so called MagneChip platform including four serial reaction chambers with individually removable permanent magnets. The results were achieved by experiments using phenylalanine ammonia-lyase (PAL) from *Petroselinum crispum* immobilized on the surface of magnetic nanoparticles (MNPs) and filled into one or more chamber of the MagneChip.

Biotransformations with PAL under different conditions were performed mostly using the natural substrate L-phenylalanine to study

- the reproducibility of biotransformation in the microreactor system,

- the effects of the long-term use and cyclic reuse of the biocatalyst on the biocatalytic activity,

- the effects of the particle size on the biocatalytic activity,

- the optimal substrate concentration and flow rate of the in-chip biotransformation,

- the effect of immobilization and the use of flow microreactor on the kinetic constants, and

- the biotransformations of further substrates with PAL.

2. Background

2.1. Microreactors

Analytical systems which comprise microreactors are characterized by outstanding repeatability and reproducibility, due to replacing batch iterative steps and discrete sample treatment by flow injection systems [4]. The possibility of performing similar analyses in parallel is an attractive feature for screening and routine use [3]. Microreactors have been integrated into automated analytical systems, as well as providing benefits from system automation, and this also eliminates errors associated with manual protocols [4].

Applications of microreactors can be divided into three classes [4]:

• Organic synthesis, when a target molecule is formed from components in flow

• Analytical use of biocatalysts to transform an analyte difficult to measure to an easy to measure form

• Screening of substrates and enzymes examines their kinetic characteristics

Microreactor systems can be further divided into classes based on the physical localization of the catalyst:

Laminar flow reactors: The majority of the commercial flow synthesis systems utilize laminar flow with soluble components and enzymes [5]. Losing the catalyst is a major drawback of this technique.

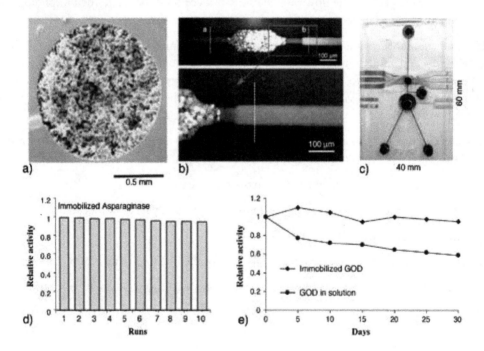

Figure 1. Lab-on-a-chip microreactors: (a) a monolith silica reactor [8], (b) a packed bed silica reactor [9], and (c) a packed bed MNP reactor [10]. Stability of immobilized enzymes: (d) after reuse of asparaginase in 10 cycles, 10 min each [7] and (e) during long-term operation of immobilized GOD [8].

Filled reactors: Microreactors utilizing immobilized catalysts (e.g., enzymes) have many advantages over the traditional flow reactors. First of all, the catalyst (enzyme) can be recycled after usage; therefore, the process is more economical and environmental friendly. The reactions are highly reproducible as the catalyst (enzyme) concentration is fixed in the system. Immobilization of enzymes often causes decrement in biocatalytic activity and choosing the appropriate immobilization technique is challenging [6]. It was reported that immobilized asparaginase retained 95.7% of its activity after 10 cycles of use [7] (**Figure 1d**), while immobilized glucose oxidase enzyme (GOD) retained 97% of its original activity after cyclic regeneration and reuse [8]. Immobilization often extends the long-term stability and temperature resistance of the enzymes, and in several cases, even the catalytic activity is increased compared with the soluble form. Immobilized asparaginase retained the 72.6% of its original activity for 10 weeks [7], and immobilized GOD retained the 95% of its activity [8] (**Figure 1e**) for 30 days.

The reactors filled with immobilized (bio)catalyst can be further divided into two groups according to the type of the supporting material of the (bio)catalyst:

- *Monolith reactors:* The reactor is defined as monolith reactor where the supporting material is fixed in the reactor volume, and the (bio)catalyst is nonremovable; therefore, the chip is single use. Nanostructured materials are used to further increase the SVR. Examples include silica monolith reactors [8] (**Figure 1a**) or most recently reactors incorporating nanofibrous material made by electrospinning [11].

- *Packed bed reactors:* The reactor is defined as packed bed reactor where the supporting materials are beads, even on microscale or nanoscale. The (bio)catalyst is immobilized onto their surfaces. Nanotechnology enables functional modifications of the beads, for example, making them magnetic. The reactor can be loaded with the suspension of the beads (**Figure 1b**) and viscous [9] or magnetic [12, 13] forces are utilized to keep the particles fixed in the reaction chamber.

Kinetic studies could be carried out with ease in microreactors by changing the attributes of the reaction, for example, the inflow substrate concentration. Because the most often used Michaelis–Menten model cannot be applied to flow reactors; in several cases [9, 14], the Lilly–Hornby model [15] was applied. Dependency of the kinetic parameters on the flow rate—and occasionally on further other parameters—was reported in many cases implying the limitations of the Michaelis–Menten model [8, 9, 14, 16].

In every on-chip study, K_m and k_{cat} as kinetic parameters were determined using various ways of product quantification such as capillary electrophoresis (CE) [7], amperometry [8], and fluorescent imaging [9, 16].

References	Enzyme	Method	Reusability	Stability	Particle	[E] measurement
Mu et al. [7]	Asparaginase	Michaelis–Menten, LB plot	10 cycles 100 min	10 weeks	MNP	Out of chip
		K_m, V_{max} CE out of chip	95.70%	72.60%	12 nm	Supernatant

References	Enzyme	Method	Reusability	Stability	Particle	[E] measurement
He et al. [8]	GOD	Michaelis–Menten, EH plot	97%	30 days	Monolith reactor	Out of chip
		K_m, k_{cat} Amperometry, on-chip		95%	silica	Absorbance
Kerby et al. [14]	Alkaline phosphatase	Lilly–Hornby	N/A	N/A	Silica Microbeads	In chip
		K_m, k_{cat} Fluorescent imaging				Optical
Slovakova et al. [16]	Trypsin	Michaelis–Menten, LB plot	80%	N/A	MNP	Approximated
		K_m, k_{cat}			600 nm	
Seong et al. [9]		Lilly–Hornby	N/A	N/A	Microbeads	Optical
		K_m Fluorescent imaging			15 μm	

Table 1. Lab-on-a-chip microreactors with immobilized enzymes.

2.2. Magnetic nanoparticles in microreactors

The importance of MNPs as potential carriers of biomolecules is growing rapidly in biotechnology and biomedicine. In LoC systems, nanosized magnetic particles provide quasi-homogeneous systems, high dispersion, high reactivity, low diffusion limits, and possibility of magnetic separation. The MNPs are usually collected in microsized reaction chambers. The collection and separation from the fluid stream are accomplished by external magnetic field. Such microreactors were found to be highly effective in biodetection [24], biocatalytic [17], and bioanalytical [18] applications (**Table 1**).

Magnetite nanoparticles exhibit superparamagnetic or soft ferromagnetic behavior with high saturation magnetization resulting in high permeability values [19]. To date, magnetic manipulation of magnetic beads utilizing a magnetic bead separator array seems to be one of the most promising technique of precise handling of biocatalysts in chip. Do et al. [12] developed a microfluidic platform, where the magnetic field was concentrated between permalloy patterns (50 × 100) to produce a high magnetic field gradient over the edges of them, thus being able to trap the magnetic beads. Li et al. [13] used external hard magnet to develop a concentrated magnetic field perpendicular to the channel at a certain position of the chip. The particles accumulated at the designated place. Slovakova et al. [16] used a pair of hard neodymium magnets positioned in a given angle to develop a magnetic field parallel to the channel structure. It was reported that in this case, the particles are arranged parallel with the channel axis, and also, the reaction efficiency was reasonably higher than in orthogonal configurations. Lien et al. [10] used an integrated electromagnet with active cooling for the entrapment of the magnetic particles in the reaction chamber (**Figure 1c**).

Because of the widespread applications of MNPs in biotechnology, biomedical, and material science, more and more synthesis techniques have been developed to obtain different kinds of MNPs. Exhaustive discussions on the available synthesis techniques (e.g., coprecipitation, microemulsion, thermal decomposition, solvothermal, sonochemical, microwave assisted, chemical vapor deposition, combustion synthesis) can be found in several reviews [20, 21]. The synthetic methods will determine the shape, the size distribution, size, the surface chemistry of the particles, and consequently their magnetic properties. Various optimization methods could be used to obtain proper MNPs suitable for the desired research and commercial applications [21].

In our study, the surface of MNPs was chemically modified by sol–gel method, which resulted in the formation of a core–shell silica-MNP carrier. Then, the surface was functionalized by epoxy groups, which were able to form stable, covalent binding with the amino, thiol, or hydroxide groups of the enzyme. The immobilization of *Pc*PAL was carried out in liquid phase. For a detailed description, see [18]. After immobilization, negligible protein contents in the supernatants of the washing procedure were determined by the Bradford assay method [22]. The resulted enzyme-coated magnetic nanoparticles (*ec*MNPs) were used in two size variations, with 250 and 600 nm diameters. Where otherwise not indicated the nominal *ec*MNP diameter is 250 nm.

3. The MagneChip platform: construction and operation

MagneChip is a microfluidic platform centered on a chip consisting of several reaction chambers enabling accumulation (and release) of MNPs. This magnetic microreactor chip can utilize the benefit of excellent separation ability of MNPs in magnetic field. In various applications, the MNPs covered by biologically active molecules (e.g., bioreceptors) are immobilized on their surfaces may be used. Magnetic techniques enable anchoring the particles inside certain compartment of microreactors, where the accumulated magnetic particles can form a dense layer. After filling (in a consecutive step), reagents can flow through the chip, while bioreaction occurs inside the microchambers and the resulted product flows through the chip. The outflow can be collected and/or quantified outside the chip, for instance by absorbance method. Because the enzyme to be immobilized on the MNP surfaces can be chosen freely, a wide variety of applications are possible (**Figure 2**). Taking the advantage of the continuous-flow operation, product formation can be monitored for a long time under various conditions over the same anchored *ec*MNP layer. MagneChip can be reinitialized periodically which enables multiparameter experiments, and therefore, reaction kinetics can be characterized in a fully automated way. Because the flow control system of the platform allows changing the actual substrate over the *ec*MNP layer, reactions can be screened even with unexplored substrates (**Scheme 1**). This feature renders MagneChip as a tool for substrate discovery as well.

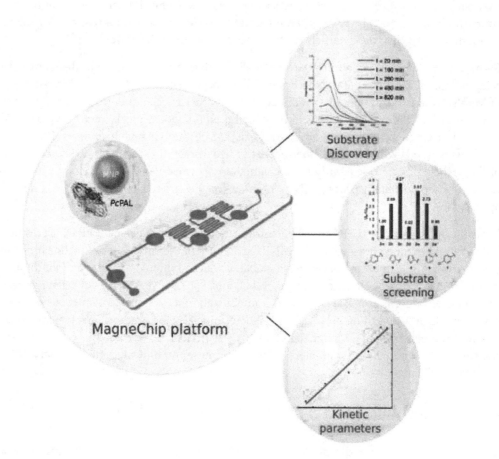

Figure 2. Possible applications of MagneChip platform. Reaction chambers are filled with bio-functionalized magnetic nanoparticles, product formation is measured by in-line UV detector.

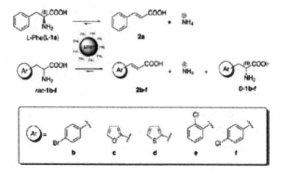

Scheme 1. Ammonia elimination from different amino acids (**1a-f**) catalyzed by *Pc*PAL immobilized onto MNPs within the MagneChip.

3.1. Basic aims and principles

A microfluidic test bench was developed for carrying out microreactor experiments with MagneChip (**Figure 3**). The test bench consisted of two syringe pumps for dispensing reagents, a thermostable chip holder and a zoom microscope for the optical inspection of the chip. The chip holder had four magnet drawers enabling to push permanent magnets under reaction chambers of the chip and also pull them out as the magnetic field is no longer required.

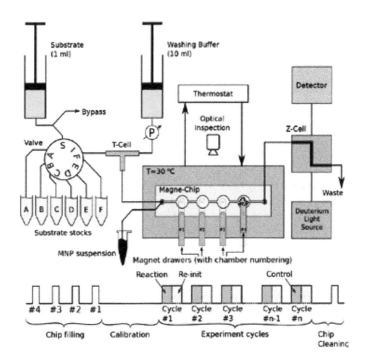

Figure 3. Schematic diagram of the fluid control system of MagneChip platform [22].

MagneChip reaction chambers (volume of ~1 μl) were designed to accomplish the following requirements:

- Because (bio)chemical reaction occurred inside the reaction chambers operating under continuous-flow conditions, a relative homogenous flow velocity distribution was required. This condition could be fulfilled because laminar flow was developed in the chamber.

- MNPs were accumulated in the chambers, and their drifting was prevented by an external magnetic field. The critical flow rate and the amount of accumulated ecMNPs could be increased using prolated channels providing sufficient amount of (bio)catalysts to reach reasonable conversion of the desired reaction.

- A resonant coil magnetometer was installed under the reaction chambers. Using the magnetometer, the accumulated amount of ecMNPs could be measured with high accuracy.

A four reaction chamber MagneChip layout is presented in **Figure 4a**. CFD simulations revealed that the flow velocity distribution inside of the chambers varied in a scale of two (**Figure 4b**). Depending on the typical flow rates used in MagneChip, reaction residence time in the chambers may vary from 1 to 10 s (**Figure 4c**).

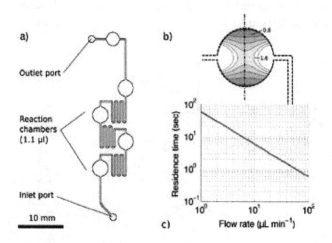

Figure 4. (a) Layout of the four chamber MagneChip; (b) flow velocity distribution inside the reaction chambers (units are in mm s^{-1}); (c) residence time vs. flow rate in MagneChip reaction chambers.

3.2. Construction method of MagneChip

Figure 4a depicts the arrangement of a four-chamber chip used for testing the enzymatic reactions.

The chip was constructed by PDMS molding technology. SU-8 photoresist structures were prepared as a molding master, resulting in a channel height of 110 μm. PDMS was poured on the master and was kept on room temperature for 1 day. After cross-linking, the PDMS replica was released and the PDMS channel bodies were bonded to standard microscope glasses after oxygen plasma treatment. Some of the chips were equipped with a resonant coil magnetometer placed under the chambers for MNP quantity measurement [23]. The coil was embedded in an intermediate PDMS layer. For further construction details, see [23].

3.3. Method of MNP quantification in the reaction chambers

The magnetic behavior of MNPs initiated the development of an inductive method to quantify the nanoparticles. The measurement is based on the resonance frequency shift of a passive electrical resonant circuit, where a flat inductor coil integrated in a silicone elastomer film acts as a sensor. From the suspension of MNPs flowing into the chip, MNPs were anchored within the reaction chambers by external permanent magnets. The *ec*MNP amount inside the chamber affected the inductance; therefore, the resonance frequency was changed. The method also enabled on-line monitoring of the actual *ec*MNP quantity in the chamber. This test arrangement enabled to study the effect of particle size and arrangement on the chamber filling MNP mass and also on the catalytic activity of the PAL bound to the *ec*MNPs [23].

3.4. Operation methods of MagneChip

3.4.1. Fluid handling steps

The experiments in MagneChip (**Figure 3**) involved four steps: (1) filling up the chip with MNPs, (2) absorbance calibration, (3) experiment cycles, and (4) chip cleaning.

Chip filling. In the chip-filling step, an MNP suspension was driven through the chip by applying a slight air pressure (0.2–0.3 bar) to the vial containing the MNP suspension and connected to the inlet of MagneChip via a PTFE tube (**Figure 3**) at 25°C. During the filling process, the MNPs were accumulated in the reaction chambers due to the permanent magnets placed in moveable drawers enabling "on/off" switching of the magnetic field. Once the chamber most distant from the inlet (**Figure 3**, *Chamber 4*) was saturated, the permanent magnet of the chamber at preceding position was turned on (**Figure 3**, *Chamber 3*). The same procedure was repeated (**Figure 3**: *Chambers 2* and *1*) until all chambers were filled up. Each chamber of the MagneChip device could capture ca. 250 µg of *ec*MNP biocatalyst [23].

Calibration and experiment cycles: During the forthcoming steps, the valve at the inlet of the MagneChip (**Figure 3**) was switched to the substrate (reagent) circuit. The flow controller performed the dosage of the substrate and other chemicals as dictated by the programmed sequence.

Chip cleaning: At the final, chip-cleaning step, the magnetic drawers of the MagneChip (**Figure 3**) were drown out and a washing solution was driven through the chip to remove *ec*MNPs.

The individual steps in a series of experiments, called experiment cycle (**Figure 3**, *Experiment cycles*), involved a *Reaction step* and a *Re-initialization step*. A series of experiments could consist of several experiment cycles depending on the number of parameters to be changed.

Reaction step. In a *Reaction step*, a substrate-containing solution was flowing through the chip at a constant flow rate, and the specific absorbance of the product was continuously monitored in the outflow of the chip.

Re-initialization step. In the *Re-initialization step,* the feed of the substrate-containing solution was stopped, and the chip was flushed with a washing buffer, while the magnetic particles were retained by the permanent magnets.

3.4.2. Reaction step variants

The substrate feed (with continuous monitoring of the absorbance in the outflow at a previously selected wavelength) was accomplished according to one of the following variants. The cycle ended when the predesigned step time had been passed or when the reaction reached saturation.

- *Repeatibility test.* The feed of the substrate started (1st cycle) or continued at unchanged flow rate.

- *Flow rate test.* The feed of the substrate started (1st cycle) or continued, while the flow rate changed cycle by cycle.

- *Substrate concentration test.* The substrate-containing solution and the washing buffer were feed in parallel at a predesigned ratio resulted in a predefined dilution of the substrate at the chip inlet. The dilution ratio could be different cycle by cycle.

- *Substrate screening.* The actual substrate was loaded into the substrate syringe through a bypass valve from the actual container of the substrate stock (A-F, in **Figure 3**), and the feed of the substrate began at a predefined flow rate. To change to the next substrate, a *Re-initialization step* was performed, followed by loading the next substrate into the substrate syringe from the substrate stock (A-F).

3.5. Quality assessment of the operations in MagneChip

A series of subsequent measurements performed by the system were considered as reliable if all the following conditions were met [22]:

- independent measurements were reproducible using the same type of *ec*MNP biocatalyst,

- the product of the enzyme reaction could be measured selectively in the UV–Vis range,

- product and substrate could be completely removed through the washing steps,

- the enzymatic activity of the *ec*MNP biocatalyst remained unchanged during the measurement

- and last but not least, the *ec*MNP layer in the magnetic reactors remained unharmed during the measurement cycles.

In order to test the fulfillment of the first group of conditions, a control measurement was performed after each series of experiments; that is, the first step of the sequence was repeated in the last step under the same conditions, and the specific activity of the immobilized biocatalyst (U_B) at saturation concentrations of L-phenylalanine (L-**1a** in **Scheme 1**) in the first and last cycles was compared.

3.5.1. Reproducibility of the individual measurements

Reproducibility of the chip-filling process: The first single chamber of the MagneChip was filled with MNP suspension. Biotransformation of L-**1a** to **2a** (**Scheme 1**) was performed in flow-through mode and monitored by in-line UV–Vis. After reaching the stationary state (i.e., constant level of product formation), the magnet of the chamber was released and the *ec*MNPs were captured in the next chamber. The experiments performed in three consecutive chambers were repeated three times resulting in $U_B = 8.01 \pm 0.14$ µmol g^{-1} min^{-1} [22].

The filling–refilling results indicated that neither that the homogeneity of the MNP suspension nor the filling procedure of the chambers had remarkable effect on the reproducibility of the measurements. The significant difference between the U_B values of MNP biocatalyst in shake vials and in MagneChip indicated increased effectivity of the biocatalysts in MagneChip device [22].

Reproducibility of independent measurements: Biotransformation of L-phenylalanine (L-**1a**) to (*E*)-cinnamic acid (**2a**) by MNP biocatalyst suspension (**Scheme 1**) was performed in shake vial as three parallel reactions and resulted in U_B = 2.91 ± 0.08 µmol g^{-1} min^{-1} ensuring that the homogeneity of the MNP suspension was sufficient [22].

3.5.2. Optical inspection of the reaction chambers

During the experiments, the chip was optically inspected by a zooming microscope and a monochrome hi-speed smart camera. Before evaluating the measurement sequence, the plan view of the chip was stored as a reference ($\prod_{ref}$). At the end of the step *i* of the measurement sequence, the plan view of the chip was sampled again ($\prod_{seq,i}$) and it was compared to the reference as follows [22]:

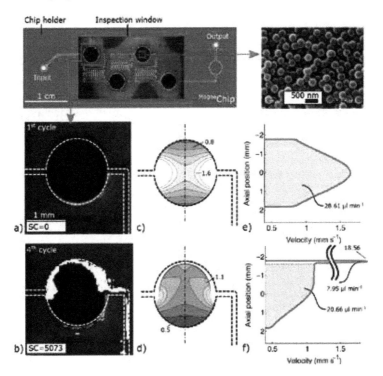

Figure 5. MagneChip device with four MNP-filled and external magnet-equipped microchambers (top left) and SEM image of the MNP layer (top right). The effect of air bubble passage through the reaction chamber [(a)–(f)]: (a) photograph, before passage; (b) difference image (difference score SC = 5073), after passage; (c) calculated flow velocity field before and (d) after the passage; (e) velocity profile in the middle cross section of the chamber before and (f) after the passage [22].

$$\Pi_{diff}(j,k) = \begin{cases} \Pi_{ref}(j,k), \Pi_{ref}(j,k) - \Pi_{seq,i}(j,k) < 0 \\ 0, \Pi_{ref}(j,k) - \Pi_{seq,i}(j,k) \geq 0 \end{cases}$$

where (j,k) are the pixel coordinates of the plan view image; therefore, the changes in accordance to the reference image are indicated by white pixels. The total number of white pixels is defined as *chamber difference score (SC)* used as a marker for describing the changes of the MNP layer arrangement. Therefore, the changes compared with the image of the first cycle (reference) were indicated by white areas during the consecutive cycles of the measurement.

In practice, SC values under 2000 reflected to negligible changes. However, $SC > 3000$ indicated serious structural change of the *ec*MNP layer, for instance, the complete breakthrough of a bubble (**Figure 5b**) [22]. Air bubbles usually did not split at the channel entrance, rather passed at one side along the chamber wall. Numerical simulations revealed (**Figure 5c–f**) that the velocity profile became asymmetric due to the bubble passage and the overall mass flow rate through the porous MNP layer significantly decreased (28.6–20.7 µL min⁻¹, roughly 72% of its original value), while the remaining fluid passed through the developed tunnel. The passing bubble could drift away particles which decreased the total mass of the biocatalyst in the reaction chamber. Therefore, the biocatalytic activity of the damaged chamber decreased and the consequent measurements were no longer reliable.

Reliability assessment of the measurements was based mostly on the following parameters:

1. Chamber difference score (SC) — Over $SC > 4000$ (average), the measurement was declined.

2. Control measurement — Over 5% of error, the measurement was declined.

Each of the experiments carried out by the platform was justified based on the above criterion.

3.5.3. Reproducibility of cyclic reactions

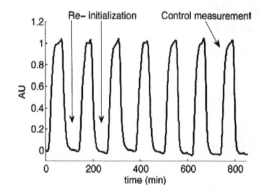

Figure 6. Time plot of the periodic absorbance change during the cyclic measurement (attempt 1, stable layer). The chip is re-initialized between the reaction steps (reaching zero absorbance) by washing out substrate and product completely [22]. The last measurement served as a control.

A crucial feature is the reproducibility of cyclic reactions performed by the system. To check the reproducibility of the test reactions, the MagneChip was filled with MNP biocatalyst, and biotransformation of L-**1a** to **2a** was performed in seven consecutive cycles, while the chip was re-initialized during the steps by washing out the substrate and product completely [22]. The absorbance plot at 290 nm in **Figure 6** with the aid of the previously measured extinction coefficient of the product (**2a**) indicated the concentration changes of **2a**.

The product quantity in cycle by cycle—calculated by taking the integral of the absorbance plot—clearly indicated that the chip was successfully re-initialized in every cycle throughout the experiment, and the reaction was repeated reproducibly seven times (average product quantity of $P = 0.12 \pm 1.5\%$ µmol) [22]. The moderate mean value of the chamber difference score $SC = 1322$ (1609 max) reflected negligible changes in the MNP layer.

4. The MagneChip platform: application examples

4.1. Characterization of the PAL reaction with L-phenylalanine (L-1a) in MagneChip

4.1.1. Influence of the substrate flow rate on the biotransformation

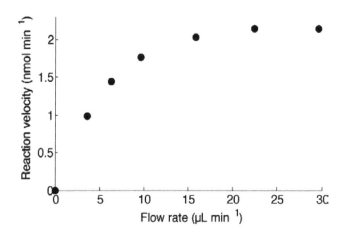

Figure 7. Dependence of the reaction rate of L-**1a** conversion to **2a** on the flow rate in MagneChip filled by *ec*MNP. Saturation was reached at 25 µL min⁻¹ [22].

MagneChip was filled with *ec*MNP biocatalyst, and biotransformations of L-**1a** to **2a** (**Scheme 1**) at various flow rates were performed in seven consecutive cycles, while the chip was re-initialized at the end of each cycle and a new substrate flow rate was set between 3.6 and 28.6 µL min⁻¹. The first (reference) measurement was repeated in the last cycle as a control. The negligible difference of specific biocatalytic activity (U_B) between the reference and control measurements (only 3%) and low SC score ($SC < 338$) indicated that the shear forces did not caused irreversible changes on the biocatalytic activity even at high flow rate (up to 28.6 µL min⁻¹) [22].

The reaction velocity was calculated for each cycles. By increasing the flow rate, the calculated reaction velocity increased until reaching saturation at about 25 μL min⁻¹ (**Figure 7**).

4.1.2. Calculation of kinetic parameters

MagneChip was filled with *ec*MNP biocatalyst, and biotransformations of L-**1a** to **2a** (**Scheme 1**) at various concentrations of L-**1a** (S_0) were performed in 10 consecutive cycles, while the chip was re-initialized at the end of each cycle and a new substrate concentration was set. It was found that the reaction followed the first-order kinetics up to (S_0) = 3 mM and saturated roughly at (S_0) = 20 mM [22].

The linear fitting method proposed by Lilly et al. [15] was applied for the calculation of the kinetic constants of the biotransformation of L-**1a** to **2a** in the MagneChip (**Figure 8**, bottom). The values of the kinetic constants are summarized in **Table 2** [22].

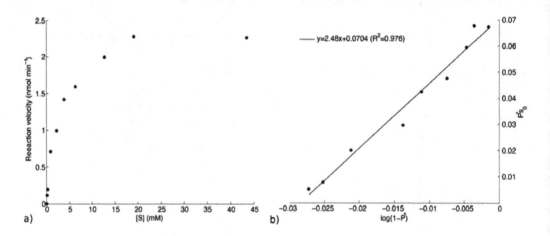

Figure 8. (a) Dependency of the substrate concentration on reaction velocity in MagneChip for the transformation of L-**1a** to **2a** by MNP biocatalyst. Saturation concentration was reached at 20 mM. (b) Linear fit based on the Lilly–Hornby model [15] to determine K_m (resulting in K_m = 2.5 mM) [22].

Kinetic parameter	MagneChip	Shake vial
K_m (mM)	2.5	9.1
k_{cat} (s⁻¹)	2.8×10^{-2}	3.2×10^{-2}
k_{cat}/K_m (s⁻¹ M⁻¹)	11.3	3.5

Tablee 2. Kinetic constants in biotransformation of L-**1a** to **2a** with MNP in shake vial and in MagneChip [22]

It was found that the apparent K_m value was reasonably smaller in MagneChip (2.5 mM) than in shake vial (9.1 mM). Turnover number (k_{cat}) and specificity constant (k_{cat}/K_m) were determined also for both reaction modes. While in the shake vial, the turnover number was somewhat higher ($3.2 \cdot \times 10^{-2}$ s⁻¹) than the in chip ($2.8 \cdot \times \cdot 10^{-2}$ s⁻¹), the specificity constant turned out to be significantly higher in chip (11.3 s^{-1} M^{-1}) as compared with the shake vial (3.5 s^{-1} M^{-1}). This may be attributed to the smaller K_m value in the MagneChip indicating significant contribution of diffusion effects to the higher apparent K_m value in shake vial.

4.2. Effect of particle size on the enzyme activity

The accumulated quantity of *ec*MNPs in the reaction chambers was determined by the embedded resonant magnetometer of the MagneChip device [23]. The measurements revealed that the total mass of the accumulated particles was approximately the same for two different particle sizes (m = 241.6 μg, *ec*MNP$_{600}$, d = 600 nm and m = 248.3 μg, *ec*MNP$_{250}$, d = 250 nm) [23]. The total particle mass could be only increased using a binary mixture (m = 283.6 μg, MNP$_{250:600}$) of the particles. This experiment resulted in a significantly higher MNP mass (17%) captured in the magnetic chamber as compared with the chamber capacity filled with MNPs of uniform particle sizes [23].

MagneChip was filled with different sized MNP biocatalysts, and biotransformations of L-**1a** to **2a** (**Scheme 1**) were also performed [23]. Compared with the larger particles (ecMNP$_{600}$), the total surface area increased both in the *ec*MNP$_{250}$ (2.5 times) and the mixture cases (2.06 times). Note that differences in biocatalytic activity can be expected only due to changes of transport limitations as the enzyme to MNP mass ratio was kept to be constant of 15% in both cases.

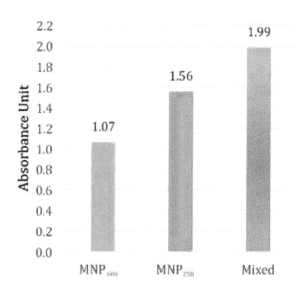

Figure 9. Specific absorbance of cinnamic acid (**2a**) at 295 nm at the chip outlet using MNP$_{600}$, MNP$_{250}$, and 1:1 mixture of the two kind of particles in the chip [25].

Results of the measurement using variously sized *ec*MNPs as biocatalysts are summarized in **Figure 9**. In fact, the *ec*MNP$_{600}$-filled chambers yielded the lowest final concentration of product as indicated by the lowest specific absorbance (AU = 1.07, at 295 nm) at the chip outlet. Filling the chip by *ec*MNP$_{250}$ resulted in an increase of the measured absorbance by 46% (AU = 1.56, at 295 nm). Because the chambers contained the same filling mass (m = 241.6 μg for *ec*MNP$_{600}$

and m = 248.3 μg for $ecMNP_{250}$) and therefore the same enzyme amount, the difference between the MNP_{250}- and the MNP_{600}-filled reactors can only be attributed to other factors, for example, to the differences in total surface area [23].

The major difference can also stem from the remarkably smaller average microchannel diameters between the particles within the $ecMNP_{250}$-filled chamber as compared to the $ecMNP_{600}$-filled one. This can result in shortened diffusion path and therefore better mass transport [23]. An additional 40% increment was achieved using the 1:1 particle mixture, which was obviously resulted as a synergy of the higher enzyme content (17%) due to the higher chamber capacity and enhanced transport phenomena due to the small average microchannel diameter [25].

4.3. Testing multiple substrates in MagneChip

Substrate screening experiments were performed with a single $ecMNP$-loading in the chip passing the solutions of the different substrates (**Scheme 1**: L-**1a** and rac-**1b-f**) through the same chip according to a predefined sequence [22]. The intensive washing procedure between the individual tests with various substrates ensured complete removal of any substrate or product from the preceding cycle (reaction). In the first cycle, the ammonia elimination was measured from L-**1a** (the natural substrate of PAL). This reaction was chosen as reference for comparison to the other elimination reactions of PAL from the further substrates (rac-**1b-f**). The difference between the initial and final (control) measurement with L-**1a** was found to be only 1.5%, while the SC score remained below 2000. Surprisingly, in the MagneChip device, higher biocatalytic activities (U_B) were observed with four of the unnatural substrates (rac-**1b,c,e,f**), than with the natural substrate L-phenylalanine L-**1a** (**Figure 10**).

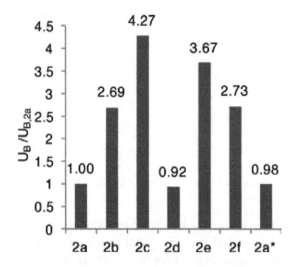

Figure 10. Comparison of the specific biocatalytic activity of $PcPAL$ immobilized on MNPs with substrates L-**1a** and rac-**1b-f** in MagneChip system [(S) = 20 mM, flow rate: 48.6 μL min^{-1}] [22]. *Control measurement.

Noteworthy, all the four unnatural substrates (*rac*-**1b,c,e,f**) which were transformed by the MNP biocatalyst with higher specific biocatalytic activity (U_B) than that of L-phenylalanine L-**1a** contained slightly more electron-withdrawing aromatic moieties than the phenyl group. This difference from the productivity ranks observed with homogenous *Pc*PAL so far may be due to the reduced contribution of the reverse reaction (equilibrium effect) to the apparent forward reaction rates in the continuous-flow system at high flow rates [22].

4.4. Characterization of an enzyme reaction with a novel substrate

By a reaction performed in the MagneChip device, it was first demonstrated that PAL can catalyze the ammonia elimination from the acyclic DL-propargylglycine (PG) to yield (*E*)-pent-2-ene-4-ynoate, indicating new opportunities to extend the MIO-enzyme toolbox toward acyclic substrates. Deamination of PG, being acyclic, cannot involve a Friedel–Crafts-type attack at an aromatic ring [18].

MagneChip, filled by PAL-*ec*MNPs, was used for the microscale biotransformation of DL-propargylglycine in sodium carbonate-buffered D_2O. The device enabled to detect the formation of (*E*)-pent-2-en-4-ynoate at 242 nm and to produce measurable quantities of the product for recording ^{1}H-NMR spectra without any work-up. Besides the significant increase of the UV-signal at 242 nm (up to A = 1.2) in the in-line UV-cell (**Figure 11**), the appearance of olefin hydrogen signals in the ^{1}H-NMR spectrum of the reaction mixture [at δ = 6.34 (*d*) and 6.85 (*d*) ppm] indicated unambiguously the formation of (*E*)-pent-2-en-4-ynoate. On the other hand, emergence of the UV signal at 274 nm during the process indicated the formation of further by-product(s) apart from (*E*)-pent-2-en-4-ynoate (**Figure 11**).

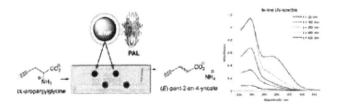

Figure 11. Ammonia elimination from DL-propargylglycine in MagneChip filled with PAL immobilized on MNPs and equipped with in-line UV–Vis detector (reaction in D_2O at pD 8.8, 37°C) [18]. The progress of the reaction was followed by full UV-spectra.

5. Conclusion

Our results proved that the MagneChip microfluidic device is a reliable, reproducible, and efficient tool which was capable of fast, reliable, and fully automated screening and kinetic characterization of *Pc*PAL substrates using minimal solvent (~500 µl) and biocatalyst (~1 mg MNP) amounts for a test compound. Compared with shake vial, the volumetric productivity of the MNP biocatalyst in the chip exceeded the one of the shake vial by more than three orders of magnitude. The platform was also capable of studying enzymatic reactions with undiscovered substrates of *Pc*PAL. The above results suggest that the MagneChip platform would be successfully utilized as a novel and flexible tool for enzyme-catalyzed biotransformations.

Author details

Ferenc Ender[1*], Diána Weiser[2] and László Poppe[2,3]

*Address all correspondence to: ender@eet.bme.hu

1 Department of Electron Devices, Budapest University of Technology and Economics, Budapest, Hungary

2 Department of Organic Chemistry and Technology, Budapest University of Technology and Economics, Budapest, Hungary

3 SynBiocat LLC, Lázár deák, Budapest, Hungary

References

[1] Ehrfeld W, Hessel V, and Haverkamp V, "Microreactors," *Ullmann's Encycl Ind Chem*, pp. 173–198, 2012.

[2] Weiler A, Junkers M. Using microreactors in chemical synthesis: batch process versus continuous flow. Pharm Technol. 2009;S6:S10–S11.

[3] Besanger TR, Chen Y, Deisingh AK, Hodgson R, Jin W, Mayer S, et al. Screening of inhibitors using enzymes entrapped in sol-gel-derived materials. Anal Chem. 2003;75:2382–91. doi:10.1021/ac026370i CCC

[4] Urban PL, Goodall DM, Bruce NC. Enzymatic microreactors in chemical analysis and kinetic studies. Biotechnol Adv. 2006;24:42–57. doi:10.1016/j.biotechadv.2005.06.001

[5] Webb D, Jamison TF. Continuous flow multi-step organic synthesis. Chem Sci. 2010;1:675. doi:10.1039/C0SC00381F

[6] Nisha S, Karthick SA, Gobi N. A review on methods, application and properties of immobilized enzyme. Chem Sci Rev Lett. 2012;1:148–55.

[7] Mu X, Qiao J, Qi L, Dong P, Ma H. Poly(2-vinyl-4,4-dimethylazlactone)-functionalized magnetic nanoparticles as carriers for enzyme immobilization and its application. ACS Appl Mater Interfaces. 2014;6:21346–54. doi:10.1021/am5063025

[8]   He P, Greenway G, Haswell SJ. Development of enzyme immobilized monolith micro-reactors integrated with microfluidic electrochemical cell for the evaluation of enzyme kinetics. Microfluid Nanofluid. 2010;8:565–73. doi:10.1007/s10404-009-0476-8

[9]   Seong GH, Heo J, Crooks RM. Measurement of enzyme kinetics using a continuous-flow microfluidic system. Anal Chem. 2003;75:3161–7. doi:10.1021/ac034155b

[10] Lien K-Y, Lee W-C, Lei H-Y, Lee G-B. Integrated reverse transcription polymerase chain reaction systems for virus detection. Biosens Bioelectron. 2007;22:1739–48. doi:10.1016/j.bios.2006.08.010

[11] Liu Y, Yang D, Yu T, Jiang X. Incorporation of electrospun nanofibrous PVDF mem-branes into a microfluidic chip assembled by PDMS and scotch tape for immunoas-says. Electrophoresis 2009;30:3269–75. doi:10.1002/elps.200900128

[12] Do J, Ahn CH. A polymer lab-on-a-chip for magnetic immunoassay with on-chip sampling and detection capabilities. Lab Chip. 2008;8:542–9. doi:10.1039/B715569G

[13] Li Y, Xu X, Yan B, Deng C, Yu W, Yang P, et al. Microchip reactor packed with metal-ion chelated magnetic silica microspheres for highly efficient proteolysis. J Proteome Res. 2007;6:2367–75. doi:10.1021/pr060558r

[14] Kerby MB, Legge RS, Tripathi A. Measurements of kinetic parameters in a microflui-dic reactor. Anal Chem. 2006;78:8273–80. doi:10.1021/ac061189l

[15] Lilly MD, Hornby WE, Crook EM. The kinetics of carboxymethylcellulose-ficin in packed beds. Biochem J. 1966;100:718–23. doi:10.1042/bj1000718

[16] Slovakova M, Minc N, Bilkova Z, Smadja C, Faigle W, Fütterer C, et al. Use of self-assembled magnetic beads for on-chip protein digestion. Lab Chip. 2005;5:935–42. doi:10.1039/B504861C

[17] Netto C, Toma H, Andrade L. Superparamagnetic nanoparticles as versatile carriers and supporting materials for enzymes. J Mol Catal B Enzym. 2013;85–86:71–92. doi:10.1016/j.molcatb.2012.08.010

[18] Weiser D, Bencze LC, Bánóczi G, Ender F, Kiss R, Kókai E, et al. Phenylalanine ammonia-lyase-catalyzed deamination of an acyclic amino acid: enzyme mechanistic studies aided by a novel microreactor filled with magnetic nanoparticles. Chem BioChem. 2015;16:2283–88. doi:10.1002/cbic.201500444

[19] Xu J, Yang H, Fu W, Du K, Sui Y, Chen J, et al. Preparation and magnetic properties of magnetite nanoparticles by sol–gel method. J Magn Magn Mater. 2007;309:307–11. doi:10.1016/j.jmmm.2006.07.037

[20] Faraji M, Yamini Y, Rezaee M. Magnetic nanoparticles: synthesis, stabilization, functionalization, characterization, and applications. J Iran Chem Soc. 2010;7:1–37. doi:10.1007/BF03245856

[21] Akbarzadeh A, Samiei M, Davaran S. Magnetic nanoparticles: preparation, physical properties and applications in biomedicine. Nanoscale Res Lett. 2012;7:144. doi: 10.1186/1556-276X-7-144

[22] Bradford MM. A rapid and sensitive method for the quantitation of microgram quantities of protein utilizing the principle of protein-dye binding. Anal Biochem. 1976;72:248–54. doi:10.1016/0003-2697(76)90527-3

[23] Ender F, Weiser D, Nagy B, Bencze LC, Paizs C, Pálovics P, et al. "Microfluidic multiple cell chip reactor filled with enzyme-coated magnetic nanoparticles — An efficient and flexible novel tool for enzyme catalyzed biotransformations," *J Flow Chem*, vol. 6, no. 1, pp. 43–52, 2016.

[24] Pamme N. On-chip bioanalysis with magnetic particles. Curr Opin Chem Biol. 2012;16:436–43. doi:10.1016/j.cbpa.2012.05.181

[25] Ender F, Weiser D, Vitéz A, Sallai G, Németh M, Poppe L. In-situ measurement of magnetic nanoparticle quantity in a microfluidic device. Microsyst Technol. 2015;21:1–12. doi:10.1007/s00542-015-2749-3

3

Cells and Organs on Chip—A Revolutionary Platform for Biomedicine

Preeti Nigam Joshi

Abstract

Lab-on-a-chip (LOC) and microfluidics are important technologies with numerous applications from drug delivery to tissue engineering. LOC integrates fluidic and electronic components on a single chip and becomes very attractive due to the possibility of their state-of-art implementation in personalized devices for the point-of-care treatments. Microfluidics is the technique that deals with small (10^{-9} to 10^{-18} L) amounts of fluids, using channels with dimensions of 10 to 100 μm. These LOC and microfluidics devices enable the development of next-generation portable and implantable bioelectronics devices. Superior chip-based technologies are emerging with the advances in microfluidics and motivating various chip-based methods for rapid low-cost analysis as compared to traditional laboratory method. An organ-on-chip (OOC) is on-chip cell culture device created with microfabrication techniques and contains continuously perfused chambers inhabited by living cells that simulate tissue- and organ-level physiology. *In vitro* models of cells, tissues and organ based on LOC devices are a major breakthrough for research in biologic systems and mechanisms. The recapitulations of cellular events in OOC devices provide them an edge over two-dimensional (2D) and three-dimensional (3D) cultures and open a gateway for their newer applications in biomedicine such as tissue engineering, drug discovery and disease modeling. In this chapter, the advancement and potential applications of OOC devices are discussed.

Keywords: lab-on-chip, MEMS, organ-on-chip, 3D cell culture, drug discovery

1. Introduction: why cell and organ on chip?

The field of microfluidics or lab-on-chip (LOC) technology aims to advance and broaden the possibilities of bioassays, cell biology and biomedical research based on the idea of miniaturization. Microfluidic systems allow more accurate modeling of physiological situations for both fundamental research and drug development [1].

Drug discovery and research is the prime aspect of any pharmaceutical company. The past 50–60 years have witnessed significant scientific and technological growth in entire field of biotechnology, computational drug design and screening and advances in scientific knowledge, such as an understanding of disease mechanisms, new drug targets and biomarkers discovery. In principal, these advancements should also be reflected in rise of new commercial products and drugs, but unfortunately, the pharmaceutical industry is facing unprecedented challenges owing to rising costs and the declining efficiency of drug research and development. Modern drug development requires implementation of extensive preclinical testing, and validation protocols before potential therapeutic compounds are approved to progress to clinical evaluation. This process is costly and time-consuming, as well as inefficient as for every 10 drugs entering clinical trials, only one or two will typically be licensed for eventual use in humans [2]. The number of new drugs approved per billion US dollars spent on R&D has halved roughly every 9 years since 1950, falling around 80-fold in inflation-adjusted terms.

The failures of drug clinical trial are primarily due to the poor predictive power of existing preclinical models. The existing cell culture techniques often failed to mimic the complexity of living systems and are incapable of modeling situations where organ-organ or tissue-tissue communication are important. Moreover, cells maintained in standard *in vitro* culture conditions often suffer from incomplete maturation or are held in a configuration that prevents their full functional development, making predictions of *in vivo* tissue function more difficult to extrapolate. Although animal models preserve the intricacy of living systems, due to the inherent complexity of interconnected tissues, elucidation of specific mode of drug action is often difficult that leads to confound observations. Furthermore, animal models have, on multiple occasions, been predicated human responses to drug treatment in a rather harmful way [3, 4]. The drug discovery community has identified the critical need for new testing approaches and an intermediate human *in vitro* model in the early stage of drug development to generate reliable predictions of drug efficacy and safety in humans that could mitigate the side effects observed in clinical trials and LOC systems can play a pivotal role in this by fulfilling this unmet need by microengineered cell culture models with miniaturized and automated assays that will increase resolution and precision. These models leverage cutting-edge microfabrication and microfluidics technologies to control the cellular microenvironment with high spatiotemporal precision and to present a variety of extracellular cues to cultured cells in a physiologically relevant context [5–6].

This chapter deals with the cutting-age research in the field of microfabrication technologies and multiorgan microdevices that mimic key aspects of human metabolism. We discuss about latest advancements and how this emerging field transforms the face of biomedicine.

1.1. Need of microfluidics technologies for global health: applications and limitations

Diagnostic applications for global health have seen a fast pace in recent years. LOC, micro total analysis systems (μ-TAS) or microfluidics systems are the major breakthrough in this regard and with their state-of-art technology, these miniaturized integrated devices have great potential to change the face of healthcare sector globally. Basically, from industrial perspective to develop a high-throughput diagnosis system, it must utilize small chemical volumes to keep the cost of development at an affordable level. The current trend of miniaturized and automated assays can address these issues directly owing to their better resolution and accuracy. Microfluidics devices are new and promising players in healthcare segments. These devices, which scaled down analytical processes in conjugation with advances in microfluidics technology, are the soul motivation behind various chip-based methods of lower cost and rapid analysis than the conventional laboratory bench-scale methods. Although these microelectro-mechanical systems (MEMS) or miniaturized chip-based systems have seen a fast pace in other fields, such as electronics, aerospace and computer science, since their inception in early 1990s and have witnessed many innovations based on these techniques, in this chapter, our prime focus is how these technological advancements have been transformed into the face of biomedical sciences with its wide range of biological applications, such as high-throughput drug screening, single cell or molecule analysis and manipulation, drug delivery and advanced therapeutics, biosensing and point-of-care diagnostics, among others. [7]

Extracting new phenomena and elaborated information about the biologically active systems is the basis of all innovations in the field of biomedical sciences. The complex live systems and richness of biological processes are stimulating factors for new LOC approaches, and these emerging technologies are gradually changing the scenario, and now, we can seek experimental answers at the molecular level.

1.1.1. Development of microfluidics technologies for different applications in healthcare segment

In a broader sense, microfluidics can be linked to the development of integrated circuit technology and wafer fabrication facilities. They have unique ability to combine different systems possessing high-throughput capabilities, new data processing and storage strategies. These miniaturized devices provide new tools for highly parallel, multiplexed assays with better isolation, purification and handling of entities, cells or organisms for a simplified, parallel analysis. Initially, silicon and related materials were the preferred choices to fabricate miniaturized devices but now polymeric materials are also the stake holders for because of ease of manufacturing by embossing or molding [8]. They are attached to other surfaces such as silicon, and the formation of fluid channels and patterns on polymeric devices are relatively easy. Other materials, such as semiconductors and metals, are other necessary components of electrical detection schemes, and earlier reports are there where semiconductor nanowires and carbon nanotubes are being studied as sensor components [9, 10]. Integration of mechanical devices with fluid systems for biological implementation and to fabricate disposable systems has been reported earlier and summarized in many reviews [7, 11, 12]. **Figure 1** shows an on-chip disposable diagnostic card. In this segment, few latest applications of LOC devices are discussed briefly [50].

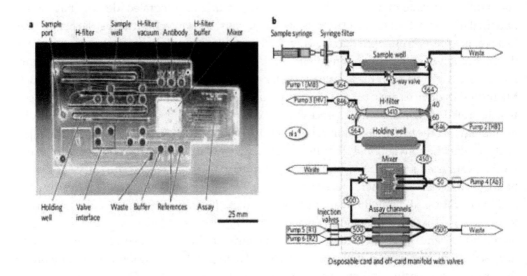

Figure 1. Example of an integrated disposable diagnostic card. (a) Image of a card. The red O-rings are for interfacing with off-card components, valves and pumps, that will eventually be incorporated onto the card itself. (b) Schematic of the card [49].

1.1.1.1. On-chip DNA hybridization and PCR

An on-chip deoxyribonucleic acid (DNA) hybridization assay refers to the bioassay conducted on the microfluidic system/device based on the nucleic acid hybridization technique [13]. From its earlier applications in 1980s, it has been evolved as a powerful tool to detect and identify the presence of a specific DNA sequence. On-chip DNA hybridization systems are amalgamation of advantages of both microfluidics and hybridization.

In the past 20 years, microfluidics devices have been emerged as an important area of research. As a combination, miniaturization eliminates the need of large reagent consumption, time-consuming labor-intensive procedures and involvement of bulky or expensive equipment while keeping its distinctive advantages of high sensitivity, selectivity and specificity of conventional techniques. Additionally, these miniaturized devices can play a pivotal role in healthcare sector of the Third World countries, by bringing cheaper and smaller, but still sophisticated analytical tools to rural areas and resource-poor regions [14]. This section focuses on few recent application of on-chip polymerase chain reaction (PCR) devices. There are few criteria to be taken care of while designing on-chip PCR systems such as high-temperature resolution and acquisition rate for precise thermal cycling in microfluidics. Apart from traditionally embedded thermocouples and thermometers [15–17], Wu et al. [18] reported an integrated PCR system with a temperature controller using platinum (Pt) thin film as heater and temperature sensor, an optical detection system and an interchangeable (disposable or modular) PCR chip, which was independent from the two functional systems as shown in **Figure 2**. In this system, Pt thin-film sensor was patterned to microsize and integrated to thin-film heater into the chip to provide rapid response and precise integration.

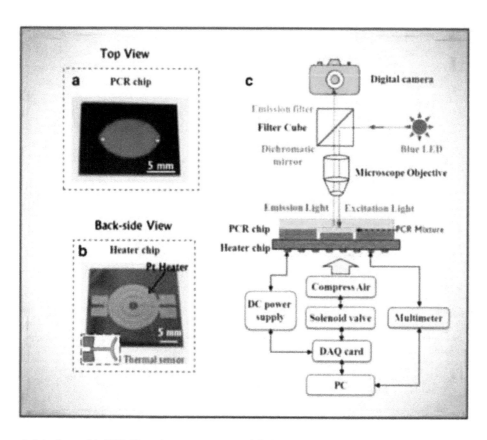

Figure 2. Interchangeable PCR chip and temperature control device. (a) Top view of PCR chip. (b) Back side view of heater chip, Pt heater and thermal sensor were integrated in one chip. (c) optical detection system in upper panel [18].

In another approach, Chia et al. developed fully integrated, portable PCR device that consists of the following four major parts: a disposable chamber chip with microchannels and pumping membranes, a heater chip with microheaters and temperature sensors, a linear array of electromagnetic actuators and a control/sensing circuit. Apart from the small size (67 × 67 × 25 mm³) and less power consumption (5V DC) and reduced volume of DNA solution, this system could effectively reduce the PCR process time into one-third of the time required by typical commercial PCR system [19]. In another approach, Steinbach et al. [20] came forward with their K-Ras mutation detection on chip. **Figure 3** shows schematic of the on-chip detection device. They aimed to develop a fast and reliable chip-based K-Ras mutation based on existing microfluidic chip platform for visual signal readout of K-Ras mutation profiling. Successful hybrid formation was monitored by streptavidin horseradish peroxidase binding, followed by an enzymatic silver deposition. Silver spots represented robust endpoint signals that enabled visual detection and grey value analysis. This study has the potential to replace expensive detection devices. These few examples give a gist of microfluidics in DNA detection and PCR. Many reviews are available on this topic [13, 21, 22].

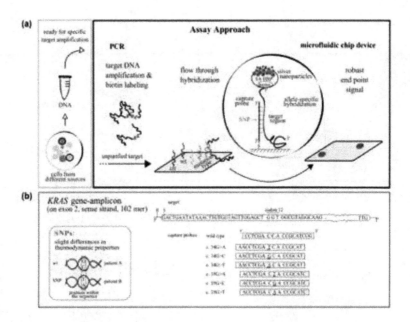

Figure 3. Assay design (a) The schematic workflow of the assay is pictured, starting from isolation of genomic DNA from cells, DNA amplification and on-chip hybridization, respectively. (b) The location of KRAS codon 12 mutations within the amplicon and the corresponding capture probes is illustrated (ctr = positive control; wt = wild type; SNP = single-nucleotide polymorphism; SA-HRP = streptavidin horseradish peroxidase) [20].

1.1.1.2. On-chip biosensing and disposable point-of-care devices

Over the past decade, on-chip diagnostic systems observed explosive growth and showed significant potential for clinical diagnostics specifically for diseases, including toxicity. The early, rapid and sensitive detection of the disease state is the prime objective for every on-chip clinical diagnosis. Initially, this field was focused on developing the concepts of LOC and later evolved to applications in a number of biochemical analysis operations, such as clinical analysis (blood gas analysis, glucose/lactate analysis, etc.) [23].

In on-chip diagnosis devices, apart from pregnancy detection kit and glucometer, most applications are based on genes and peptides detection for early indicators of disease [24–26]. For instance, Dinh et al describe a multifunctional biochip with nucleic acid and antibody probe receptors specific to the gene fragments of *Bacillus anthracis* and *Escherichia coli*, respectively [25]. These devices were based on the detection of specific diseases or biological warfare agents by incorporating biomarkers specific to such agents. Monitoring of regular metabolic parameters, such as glucose and lactate, was demonstrated by the I-Stat analyzer that provides point-of-care testing for monitoring a variety of clinically relevant parameters [26]. Immunosensing applications as a part of clinical diagnostics have also been demonstrated [27, 28].

Recent years have witnessed a vast range of applications of LOC due to the significant benefits of small sample and reagent volume utilization, economic and rapid analysis with less wastage and possibility of developing disposable devices. Ahn et al. demonstrated a fully integrated module of wristwatch-sized analyzer that included a smart passive microfluidic manipulation system based on the structurally programmable microfluidic system (sPROMs) technology, for preprogrammed sets of microfluidic sequencing with an on-chip pressure source for fluid driving, sequencing and biochemical sensors [23]. Point-of-care testing (POCT) is one of the most impressive developments of microfluidics in life sciences and can be defined as diagnostic testing at or near the site of patient care to make the test convenient and immediate. In many countries, DNA test kits for HIV are already available [29]. This is a rapidly growing field, and more detailed information can be obtained from various reviews in this area [23, 31, 32].

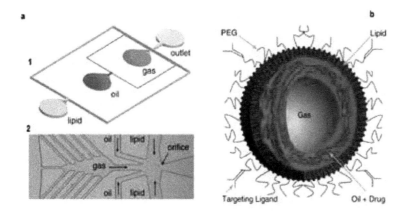

Figure 4. (a) View of microfluidic chip featuring the two distinct hydrodynamic flow-focusing regions and expanding nozzle geometry with a narrow orifice. All channels have a rectangular cross section and a height of 25 μm. (b) View of targeted lipospheres with gas in core and active ingredient in lipid oil complex [36].

1.1.1.3. Drug delivery applications

The major objective of drug delivery systems is to localize the pharmacological activity of the drug at the site of action as targeted drug delivery systems directly deliver the payload to the desired site of action with minimum interaction with normal cells. This phenomenon is especially important for anticancer drugs, as their toxicity to healthy cells is a cause of concern to improve therapeutic response and patient compliance. Last decade witnessed tremendous growth in targeted dosage forms for controlled release [33–35].

Approximately 10 million people suffer from different kinds of cancer per year and many of them unfortunately die due to lack of better treatment strategies. With the advancement in diagnostic, therapy techniques and nanomedicine, now better understanding of disease onset and treatment is possible, but still more will be offered by state-of-art microfluidic technology in terms of control over particle size, composition, encapsulation rate and better performance of nanoformulations, which have a great impact on the cancer survival rate.

In the series of microfluidics-based delivery systems, a gas-filled lipospheres was reported by Hettiarachchi et al. for targeted delivery of doxorubicin, using polydimethylsiloxane (PDMS)-based microfluidic chip that contained two distinct hydrodynamic flow-focusing regions for local administration into tumor tissues as shown in **Figure 4a** [36]. Generally, liposomal-encapsulated doxorubicin suffers from relatively nonspecific biodistribution due to size selection and nontargeted accumulation [37]. As a solution, Hettiarachchi et al. prepared multilayer liposheres with oil layer of triacetin (capable of carrying bioactive molecule) sandwiched between inner gas-filled core and outer lipid layer (polyethylene glycol (PEG) lipid conjugate DSPE-PEG2000-Biotin) with avidin as targeting moieties based on the fact that multilayer gas-filled liposheres for high payload delivery at target sites could overcome the limitations of liposomal preparation. **Figure 4b** is representation of the modified delivery system.

Another strategy that is gaining importance in diagnosis and treatment of cancer is theranostic nanomedicine that combines imaging, diagnostic agent and antitumor agent. Theranostic lipid complex nanoparticles formed by bulk mixing do not give control over composition and size which can be overcome with a microfluidic setup [37, 38]. A static micromixer-coaxial electrospray (MCE) for the single-step synthesis of theranostic-lipid complex nanoparticles (cationic lipid-nucleic acid complexes called lipoplexes) was designed by Wu et al. to overcome this limitation. Multicriteria evaluation (MCE) technique produced monodispersed particles with a diameter of ~194 nm and high encapsulation efficiency compared to a more conventional bulk process; the advantage of this process is shown in **Figure 5a**. Quantum dots (QD605) and Cy5-labeled antisense oligodeoxynucleotides (Cy5-G3139) were encapsulated as the model imaging reagent and therapeutic drug, respectively, with successful cytoplasm to delivery of drug into cytoplasm of A549 cells (nonsmall-cell lung cancer cell line) leading to $48 \pm 6\%$ down regulation of the Bcl-2 gene expression [37].

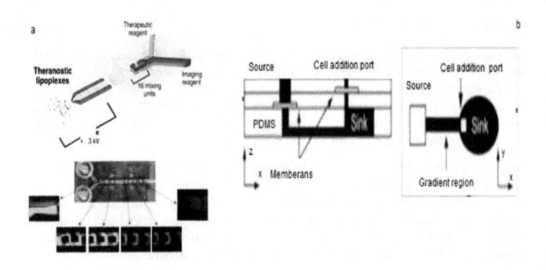

Figure 5. (a) Schematic drawing of the static micromixer-coaxial electrospray (MCE) showing its various components [37]. (b) Schematic of microfluidic gradient generator [40].

A *microfluidic gradient generator* (MGG) was developed by Abhyankar et al. [39, 40] for testing drug response on a cellular basis. These devices offered unique features of, higher resolution, real-time observation, tunable drug concentration and reduced costs in comparison with their conventional counterparts, Transwell and Dunn chambers. MGGs are based on two techniques —gradient achievement through time-evolving diffusion or parallel streams mixing. **Figure 5b** shows a sink-source flow-free gradient generator. The absence of convection flow is the key advantage of this system that eliminates the shear-stress induced to cells.

Apart from nano-based drug delivery techniques, administration of drug to the whole body is another application of microfluidics where miniaturized needles can be designed (*micronee-dles*) for improved delivery effectiveness and reduce the pain related to drug administering.

Microneedles can be classified into the following four general types: (i) solid microneedles, (ii) drug coated, (iii) polymeric microneedles with encapsulate drug that fully dissolve in the skin and (iv) hollow microneedles for drug infusion into skin.

1.1.1.4. Microarrays technologies

A microarray is an analytical device that comprises an array of molecules (oligonucleotides, cDNAs, clones, PCR products, polypeptides, antibodies and others) or tissue sections immobilized at discrete ordered [41]. In a general microarray device, sample solutions are confined in microfabricated channels and flow through the probe microarray area. Enhanced sensitivity is obtained due to high surface-to-volume ratio in microchannels of nanoliter volume and advantages of both fields can be exploited simultaneously by combining DNA microarray with microfluidics [42, 43]. Consumption of small volumes in microfluidic systems is an added advantage to develop low-cost, compact and portable LOC systems. Secondly, the surface hybridization of target DNA can also be accelerated on microfluidics platform by electrokinetic delivery of negative charged DNA molecules on to the probe area [44].

Lee et al. proposed a recirculating microfluidic device for the hybridization of oligonucleotides to DNA microarray [45]. Peristaltic pump was connected to the both ends of the microchamber to generate circulatory flow as shown in **Figure 6a**. With this device, hybridization time was also shortened to 2 h and sample volume was 100 μL.

Many companies are involved in designing microfluidic technology for various high-throughput applications, such as immunoassays, diagnostic devices, single molecule DNA and protein detection as well [42]. Researchers from the University of Chicago, USA, and other laboratories demonstrated the use of two-phase droplet systems that generate droplets within microfluidic channels to be used as microreactors for high-throughput screening of compounds and multiple chemical reactions [46, 47]. Recently, Huang et al. presented a microfluidic device integrated with pneumatically controlled microvalves and micropumps for parallel DNA hybridizations to analyze 48 different DNA targets (18-mer oligonucleotides derived from the Dengue viral genes) simultaneously. A schematic of device is shown in **Figure 6b** [48].

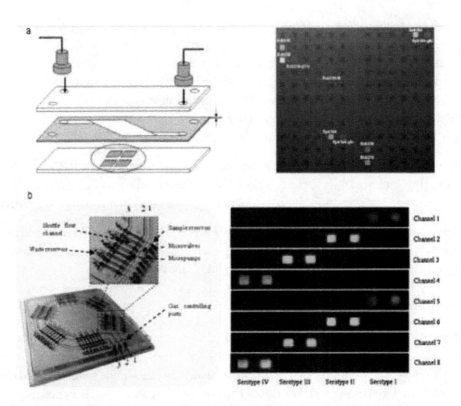

Figure 6. (a) Diagram for sample recirculation system on the hybridization chamber and hybridization image of fluorescence-labeled target nucleotide [45]. (b) *Left:* Photograph of the microfluidic chip containing shuttle-flow channels, microvalves and micropumps. The shuttle flow hybridization was realized by controlling the gas ports 1, 2 and 3 automatically. *Right:* Hybridization specificity assay using four serotypes of Dengue virus under shuttle flow conditions (frequency 2 Hz) in channels. The duration of hybridization process was 90 s and washing time was 30 s [48].

The commercialization of microarray and microfluidic technologies is evolving very fast as demonstrated by the emergence of many start-up companies due to its state-of-art technology. Affymetrix is an example where they generated a new market based on their GeneChip® technology over a 12-year period.

1.1.2. Challenges for lab-on-chip devices

Apparently, microfluidics devices have the potential to serve different scientific needs of healthcare and biomedical sectors and as we discussed earlier, their several successful applications have already been reported. The major advantages associated with miniaturized systems are faster/more accurate diagnoses; better epidemiological data for disease modeling; vaccine introduction; and utilization of minimally trained healthcare workers and better use of existing therapeutics but still many hurdles are there in broader applications of microfluidics systems.

However, there is always a silver lining and due to vastly increased interest in global health issues, the current funding climate for the development of diagnostics kits is significantly good.

Financial support for new and improved diagnostic tools for priority diseases, such as tuberculosis and cancer, is there. The Gates Foundation's Grand Challenges in Global Health initiative is supporting the development of prototypes of a disposable/hand-held reader system [49]. Thanks to increased attention on the global health issues and the motivation for their better treatment, we are witnessing the beginning of microfluidics diagnostic devices for early detection of these fatal diseases in coming few years.

We started our discussion on the issues of need of miniaturized devices for pharma industry and biomedicine. After a brief overview on impact of existing LOC systems on global health, we discuss how the new emerging cells and OOC techniques will have an everlasting effect on different areas of human health. The latest progress in microfluidics has led to the development of OOC microdevices, which recapitulate the complex structure, microenvironment and physiological functionality of living human organs. The practical implementation of these miniature organ systems is revolutionary for the field of biomedical sciences and will play a pivotal role for drug discovery and will improve our understanding for mode of action of molecules of therapeutic potential—overall, this state-of-art technology is expected to be a boon for pharma and healthcare sector.

2. Evolution of cells and organ on chip: from 3D culture to organ on chip

The process of growing eukaryotic cells *in vitro* was put forth by Harrison in 1907 to investigate the origin of nerve fibers [50] and since then its almost 100 years, these 2D cell cultures have greatly advanced our knowledge of cellular biology. They have been routinely and diligently undertaken in thousands of laboratories worldwide. However, the 2D cell cultures are arguably primitive and do not reflect the anatomy or physiology of a cell or tissue microenvironment in true sense. Two-dimensional (2D) cell cultures oversimplify the extracellular matrix (ECM) and cell microenvironment and the processes, such as drug delivery, toxicological analysis, gene expression and apoptosis, may not be directly taken up for the *in vivo* experiments from 2D analysis as ECM is completely different in *in vitro* and *in vivo* and cannot be adequately mimicked by 2D cell systems [51, 52]. These limitations of 2D cell culture led to the innovation of 3D cell culture methodologies; the concept that gave birth to the idea of OOC devices. In 3D culture, cells are grown in extracellular matrix, that is, hydrogels, scaffolds or on hanging drops. The cells, growing in third dimension, exhibit enhanced expression of differentiated functions and improved tissue organization but require a multidisciplinary approach and expertise [53, 54].

Generally, spheroids, cell aggregates and cell sheets are the common platforms for 3D culturing [55–60]. Basic objectives for developing 3D cell culture systems vary from engineering tissues for clinical delivery to the development of models for drug screening. It was observed that certain cellular processes of differentiation and morphogenesis for tissue engineering occurred preferentially in 3D instead of 2D.

In one study by Slamon et al., alteration of cellular architecture between 2D and 3D cells was observed in the growth of SKBR-3 cells that overexpress HER2, an oncogene found to be overexpressed in approximately 25% of breast tumors [61]. Cells grown as 3D spheroids using p-HEMA-coated plates had HER2 homodimers form, while in 2D cultures, HER2 formed heterodimers with HER3 [61]. Recently, Choi et al. [62] also reported that human neural stem cells with familial Alzheimer's disease mutations when grown in 3D culture recapitulate both amyloid-β plaques and neurofibrillary tangles. 3D cell culture more accurately simulates normal cell morphology, proliferation, differentiation and migrations. Similarly, in chemotherapy procedures, a difference in sensitivity to drug exposure was observed in cells grown in 2D or 3D microenvironments [63]. A study by Tung et al. indicated that A431.H9 cells grown in 2D and 3D show differences in viability when treated with the same concentrations of 5-fluorouracil (5-FU) and tirapazamine (TPZ). In the case of 5-FU, 2D cultures were reduced to approximately 5% viability following a 96-h treatment (5-FU; 10 mM), whereas 3D cells treated with the same concentration and duration, showed 75% viability; indicating that these 3D spheroids were more resistant to the antiproliferative effects of 5-FU [64].

In recent years, an increasing shift in research focus from 2D cells cultures to 3D cell cultures occurred which in turn translated 2D *in vitro* research to 3D *in vivo* animal models.

2.1. Advantages and limitations of 3D cell culture

- Flexible synthesis approach in 3D cell culture allows facile manipulations for cellular microenvironment modeling.

- With 3D cell culture systems, study at different states of disease models can be done in a similar tissue microenvironment that may reduce the need of animal testing.

- 3D culturing is more authentic way of monitoring drug metabolism studies instead of 2D. Due to the presence of layers of cells in 3D culture with tightly bind cells as compare to a monolayer in 2D, drug diffusion to cells by blocking or slowing simulate the real barriers for drug action.

- Scaffolds to support 3D cell with simultaneous growth factor, drug or gene delivery can also be synthesized.

- 3D cell culture has direct applications in tissue engineering and regenerative medicine.

Figure 7 is schematic of various methods of synthesis of 3D culture, including hanging drop, forced floating method, etc.

It is an evolving field and requires further research for its optimization, and therefore, it is evident that some clarity is needed in selecting the best method for the generation of 3D cells from individual cell lines. Additionally, the best established 3D culture methods currently available produce avascular tumor models that failed to mimic the full architecture of *in vivo* tissues and vascularization aspect of tumor development is left out, which is a huge significant part of true tumorigenesis. These limitations are the prime hurdles in the application of 3D culture as potential drug discovery tools.

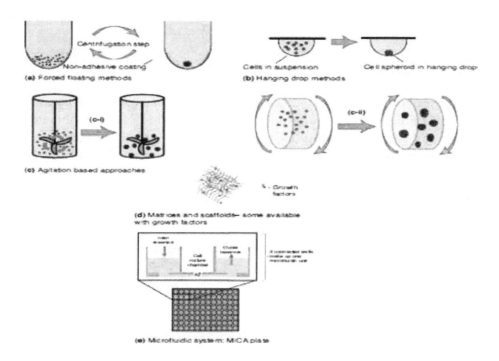

Figure 7. Schematic of 3D culture synthesis methods. These methods include forced-floating of cells; hanging drop methods; agitation-based approaches; the use of matrices or scaffolds; and microfluidic systems [53].

2.2. From 3D culture toorgans on chips: a giant leap toward biomedicine revolution

In previous section, we discussed the role of 3D cell culture and its significant impact on different fields. The next important step of 3D microfabrication is evolution of integrated OOC microsystems with the ability to mimic key structural, functional, biochemical and mechanical features as well as interactional effect of microenvironment on cell and tissues *in vivo* of living organs in a single device [65]. By definition, OOC devices are microfluidic devices for culturing living cells in continuously perfused, micrometersized chambers in order to model physiological functions of tissues and organs [66].

Cellular behavior and its interaction with *in vivo* microenvironment is still an unsolved mystery. Advancements in the field of 3D OOC opened entirely new possibilities to create *in vitro* models that reconstitute more complex, 3D, organ-level structures, with integrated chemical signals and important dynamic mechanical cues. OOC devices not only mimic the cells biomechanical and biochemical behavior in *in vivo* tissue but also predict the interactional effects of microenvironment on cells and tissue functions [58]. This unique ability of OOC devices makes them a potential candidate for drug discovery programs and a boon for healthcare segment. Though this state-of-art innovation is in its nascent state, preliminary data obtained had shown promising future of OOC devices with wide applications in biomedical sciences. As a proof of concept, researchers have fabricated two stacked PDMS cell culture chambers separated by permeable synthetic membrane to study polarized functions of various epithelial cells of intestine [67, 68], lung [69], kidney [70], heart [71], etc.

2.3. Basic microfabrication techniques and material for OOC devices

To mimic *in vivo* organ-specific microenvironment, OOC devices required high precision and accuracy. Microfabrication techniques are the preferred methodologies to fabricate OOC devices due to feasibility of constructing tissue-specific environment at microscale. Typical techniques include replica modeling, soft lithography and microcontact printing [52, 66, 72]. **Figure 8** is a schematic representation of these techniques.

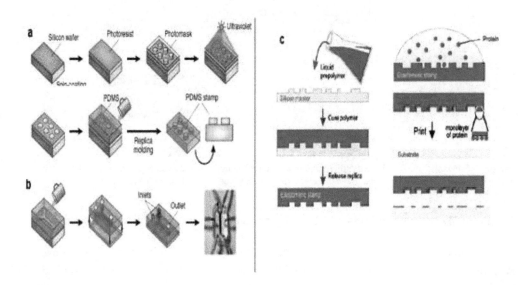

Figure 8. Schematic of microfabrication techniques. (a) Replica modeling. (b) PDMS stamp for formation of microchannels [158]. (c) Microcontact printed protein for cell pattering [159].

Replica molding techniques have been used to replicate complex surface relief patterns to produce biomimetic structures that mimic organ-specific microarchitecture. Lee et al. designed the replica modeling techniques to recreate the artificial liver sinusoid and natural endothelial barrier layer in liver. [73] This was an important breakthrough that successfully reconstituted a tissue-tissue interface that was a critical element of whole liver organ structure, and was not possible in conventional 3D ECM gel cultures. In other report by Esch et al [74], photolithography was explored to recreate the key aspects of villi structure on microfluidic chambers covered by 3D shaped, porous membranes for models of the gastrointestinal tract epithelium by two-exposure step fabrication process. As shown in **Figure 9**, complete crosslinking was used to fabricate the chamber and partial with SU-8 to form the porous membrane. This microdevice could create better *in vitro* models of human barrier tissues, such as the gastrointestinal tract epithelium, the lung epithelium or other barrier tissues with multiorgan "body-on-a-chip" devices for drug-screening application.

An array of PDMS microchambers interconnected by 1 µm wide channels was similarly used to enable growth and *in vivo*-like reorganization of osteocytes in a 3D environment that replicated the lacuna-canalicular network of bone [76]. In a similar approach, Sudo et al. came up with the idea of a microdevice incorporating ECM gels microinjected between two parallel microchannels to investigate vascularization of liver tissues in 3D culture microenvironments [76], while a compartmentalized microfluidic system for coculturing of neurons and oligodendrocytes to study neuron-glia communication during development of the central nervous system was developed by Park et al. [77]

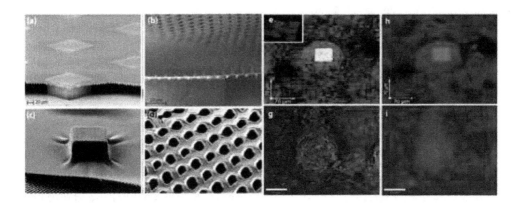

Figure 9. Porous SU-8 membranes that are anchored to and span across microfluidic chambers. The membranes are either flat (a and b), or they were dried over sacrificial silicon pillars and take on the shape of the pillars (c and d). (b) A higher magnification scanning electron microscopy image of a flat membrane with 3 µm pores. (d) Close-up of the 3D-shaped membrane imaged in (c). The image reveals the membrane's porous character. The sacrificial silicon pillars can be removed via xenon difluoride etching 3D cell culture of gastrointestinal epithelial cells (Caco-2) that were grown for 8 days (a, b, c) and 21 days (d, e, f) on porous SU-8 membranes that were dried on silicon pillars (50 µm wide and 200 µm high) [74].

From their inception, production of these microdevices relied on silicon microfabrication and micromachining techniques. Although widely explored and applied, silicon micromachining is rather complex, costly with limited accessibility to specialized engineers. To overcome these practical hurdles, researchers developed microfluidic systems made of the silicone rubber, poly(dimethylsiloxane) (PDMS), that are less expensive and easier to fabricate, which opened entirely new avenues of exploration in cell biology. [6]

PDMS has several unique properties that make it a perfect choice for the fabrication of microdevices for the culture of cells and tissues. First, PDMS possesses superior gas permeability and flexibility for adequate oxygen supply to cells in microchannels, which eliminates the need for separate oxygenators, commonly required in silicon, glass and plastic device and is particularly important to maintain differentiated function of primary cells of high metabolic demand [54, 78]. PDMS microfluidic systems enabled the formation of viable and functional human tissues.

Excellent optical transparency is prime advantage of PDMS that enabled real-time monitoring of nitric oxide production and variation in pulmonary vascular resistance in a microfluidic model and cell morphology, tissue repair and reorganization. [79–81]

Moreover, control of cellular parameters is another important phenomenon in designing OOC devices and recent advances in microfabrication techniques have significant contribution toward efficient monitoring and control of cellular responses and study of broad array of physiological factors that wasn't possible with 3D static cultures. Electrical, chemical, mechanical and optical probes for direct visualization and quantitative analysis of cellular biochemistry, gene expression, structure and mechanical responses also can be integrated into virtually any microfabricated cell culture devices and more relevant data can be obtained with these advanced OOC devices. [54, 66]

3. Organ-on-chip devices: concept to application

In this section, various state-of-art existing OOC platforms and their structural features, working principles, potential and feasibility for biomedical application are discussed. OOC devices can be defined as microfluidics systems for living cells culturing in continuously perfused, micrometersized chambers in order to model physiological functions of tissues and organs [66]. The prime objective of this emerging technique is to fabricate minimal functional units of an organ that recapitulate tissue- and organ-level interactions. These devices have great potential for investigating basic mechanisms of organ physiology and are well suited for the study of biological phenomena that depends on tissue microarchitecture and perfusion and last for relatively short span (< I month). These chips often consist of featuring multiple, controllable parallel channels, splitting and merging channels, various pumps, valves and integrated electrical and biochemical sensors. Some kind of microenvironment stimuli derived from organ-level functions can be applied to cells from certain organ.

3.1. Basic working mechanism of OOC devices

OOC systems are basically elaborated microengineered physiological systems that reconstitute the key features of specific human tissues and organs and their interactions as depicted in **Figure 10** [82, 83].

Key factors in OOC designing include the following:

- Fabrication of OOC devices start with identifying the key aspects of biochemical, mechanical environment of specific organ, including local factors from neighboring cells or tissues and stretch of organ. [82].

- The final step is to measure the functional output parameters of the cultured cells.

Earlier, with 2D and 3D cell cultures, efforts were taken to control and regulate the cell growth, shape and other cellular events but due to lack of precise 3D environment, these models suffered with inaccuracy and reliability in recapitulating the issue- and organ-specific systems

[83]. But with the state-of-art OOC technology, new possibilities to create efficient *in vitro* models with organ-specific microenvironments, tissue microarchitecture reconstruction, spatio-temporalchemical gradients, tissue-specific interfaces, crucial dynamic mechanical cues and biochemical signals [54, 84]. In this section, we describe recent progress in this field and currently reported OOC devices such as liver, kidney, intestine, kidney, heart, skin and blood vessels.

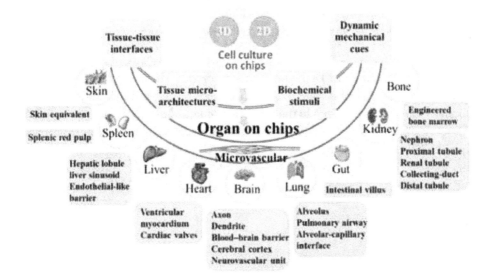

Figure 10. Representation of organ-on-chip device and concept of modeling, a complex microenvironment and their existing simulation of functional units [82].

OOC devices can be classified into three broad segments based on the working mechanisms: [82]

i.   Membrane-based penetration and mechanical stimuli—blood-brain barrier, lung, kidney, gut, heart on chip.

ii.   Organ function mimicking based on anatomy—arteries and spleen on chip.

iii. Perfusion-based OOC devices—liver, brain and womb on chip.

3.2. Membrane-based organ on-chip devices

To study the drug response with respect to human biological barriers is a crucial step in drug discovery. Researchers developed 3D compartmentalization with membrane-based multilayer compartments for mimicking biological barriers such as the blood-brain barrier [85, 86, 99], the kidney transport barrier [87, 71], and the lung's alveolar-capillary interface [88, 89] that can be considered a major breakthrough for biomedicine. In this segment, recent discoveries in membrane and muscular thin films to recapitulate the physiochemical interface and mechanical cues are described.

3.2.1. Lung on a chip

Lung is an important organ of respiratory system for the exchange of oxygen and carbon dioxide in blood stream. The elementary tissue unit of the lung is the layer of epithelial and endothelial cells over which the exchange of gases between air and blood takes place. The geometry of the lung tissue contains the epithelial-endothelial interface, epithelium-air interface, endothelium-blood interface and periodic mechanical force with each respiratory cycle. Understanding of cell-cell interactions, cell-blood and cell-gas flow is utmost necessary for drug discoveries and physiochemical research. Complex geometric and compositional structure of lung is the great barrier to enable straightforward manipulation and observation of cells.

Lung-on-chip is the microreplica of the lung on a microchip. This is used for nanotoxicology studies of various nanoparticles that are introduced into the air channels and to understand the pulmonary diseases where due to the formation of liquid plug that blocks small airways and obstruct gas flow in alveoli [89]. To understand the mechanism of liquid plug propagation and rupture, Huh et al. designed a microengineered system that consists of two PDMS chambers separated by thin polyester membrane with 400-nm pores. This system mimicked an *in vivo* basement membrane for small airway epithelial cells (SAECs) attachment and growth.

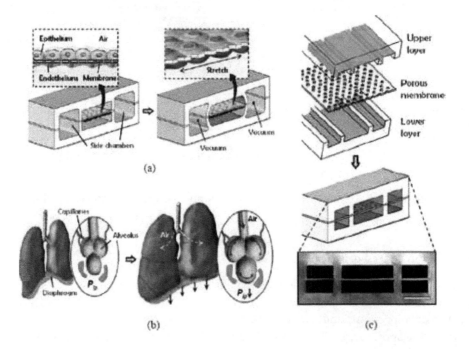

Figure 11. Schematic of lung on-chip system. (a) PDMS-based membrane to mimic alveolar capillary barrier and a vacuum based deformation controller. (b) Size variation of lung during inhalation. (c) Bonding and alignment of three layer PDMS devices [70].

Using this system, injurious response of SAECs to propagation and rupture of finite liquid plugs at an air-liquid interface afflicted with surfactant deficiency was demonstrated [88]. Another report by Huh et al. designed an alveoli-on-chip having alveolar and the capillary interface. To mimic the breathing pattern, two chambers were constructed at the side through which air is pumped in at certain required pressure, continuous increase and decrease of the flow is done in order to accomplish the inhalation and exhalation pattern. A thin flexible layer of PDMS was used in the central chamber where coculturing of human alveolar epithelial cells and blood vessel wall cells on the opposite sides is done. The membrane stretches and relaxes according to the flow of air. The culture medium is pumped through the lower microchannel to mimic the blood flow and the sample is injected on the top layer that interacts with the alveolar epithelial cells as shown in **Figure 11** [70]. In another model to study alveolar cell complexities, Douville et al. put forth their system consisting of two compartments—alveolar chamber and actunation channel. These chambers were separated by a PDMS thin membrane to create both cyclic stretch and fluid mechanical stresses. This *in vitro* model successfully demonstrated the difference in morphological changes cells undergo when exposed to combine stresses as compared to cells exposed solely to cyclic stretch [90].

These inventions reconstituted the critical lung functions and can be applied for *in vivo* models in environmental toxins, absorption of aerosolized therapeutics and the safety and efficacy of new drugs. Such a tool may help accelerate pharmaceutical development by reducing the reliance on current models, in which testing a single substance can cost more than $2 million [54, 66].

3.2.2. Kidney on chip

The word kidney-on-chip suggests that the kidney is mimicked on a chip. Here, the renal cells or the nephrons are mimicked on the chip and this is used for checking the toxicity of drug and its screening. This model helped to know more about the filtration, reabsorption of the necessary molecules from the drug as kidney toxicity is a cause of concern during drug development [91]. Nephron is the basic unit of kidney and mainly consists of glomerulus, which acts as a filtering unit that helps in filtering unwanted toxic particle from the required molecules and helps in throwing out these unwanted molecules. Nephron's glomerulus, proximal convoluted tubule and loop of Henle are mimicked on the chip. As reported by Weinberg et al., an artificial nephron function with three components on a single chip was designed [92]. Jang at al developed an on-chip kidney to reproduce cisplatin nephrotoxicity. Their device contained two compartments, where top channel mimicked urinary lumen and has fluid flow, whereas the bottom chamber imitate interstitial space filled with media. Kidney cells have less shear stress than endothelial or lung cells. This device was operated with 1 dyn/cm^2 of sheer stress [93]. A modified version of same device using human proximal tubular cells was also developed by the same group. The advantage of using proximal cells was there less sheer stress ~0.2 dyn/cm^2 that is similar to that of the living kidney tubules surrounding as shown in **Figure 12** [94]. Better understating of filtration pattern and absorption behavior that leads to toxicity was the prime aspect of this discovery.

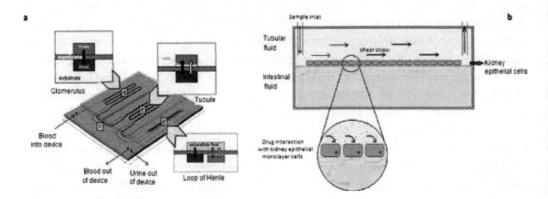

Figure 12. (a) Nephron on a chip: Schematic of the chip with cross sections of three functional units named glomerulus, proximal convoluted tubule and loop of Henle, which are response for filtration, reabsorption and urea concentration, respectively. (b) Kidney reabsorption functions using a microfluidic chip comprising of an apical channel separated from a bottom channel by proximal tubular epithelial cells cultured ECM-coated porous membrane [94].

3.2.3. Blood-brain barrier on chip

To understand and treat neurological diseases, proper understanding of blood-brain barrier (BBB) is utmost important. By definition, BBB is a unique selective barrier membrane that obstructs the passage of most exogenous compounds in blood to the central nervous system (CNS) while permeable for essential amino acids and nutrients. It is made primarily of three different cells: endothelial, pericytes and astrocytes, and the membrane is formed by firm junctions between endothelial cells that control compound permeability with high values of transendothelial electrical resistance (TEER) [82, 95]. Hatherell et al. designed a membrane-based system to replicate BBB by cultivating endothelial cells on the top side of a transwell membrane while cultivating astrocytes with or without pericytes on the opposite side [88]. However, due to low porosity and uneven pore distribution, this artificial membrane failed to recreate the close proximity to cell interaction. To address this issue, silicon nitride membrane was developed by Ma et al. to increase the direct contact between the cells and astrocytes [96]. Another report by Shayan et al. also demonstrated a considerable reduction of the flow resistance across a nanofabricated membrane with controlled pore size and low thickness (3 μm) and maintenance of metabolic activity and viability for at least 3 days [97]. A novel BBB *in vitro* model was developed by Brown et al. for efficient cell-to-cell communication between endothelial cells, pericytes, and astrocytes and independent perfusion with vascular chamber and brain chamber separated by a porous membrane (**Figure 13a**) [98]. Booth and Kim also developed a BBB that impersonated the dynamic cerebrovascular environment having fluid shear stress and a comparatively thin culture membrane of 10 μm (**Figure 13b**). This system has two components called lumenal and ablumenal on which endothelial and astrocytes were cultured to form the neurovascular unit [99].

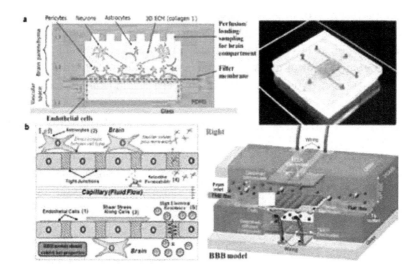

Figure 13. Schematic view of the neurovascular unit (NVU) indicating major components, cell types and their spatial arrangement. (a) Illustration of key properties should be included in an effective *in vitro* microfluidic blood-brain barrier (μBBB) models (left). (b) Structure of microdevice consisting of two channels for astrocytes and endothelial cells culture with electrodes for transendothelial electrical resistance (TEER) measurement [98, 99].

These novel systems are the promising tools of future due to their unique characteristics of feasible real time, TEER and selective permeability to study barrier function and delivery of drugs to CNS.

3.2.4. Heart on chip

Heart on chip was developed to imitate the contractility and electrophysiological response of heart in *in vitro* condition. Microfluidics has previous applications *in vitro* on cardiomyocytes, which generates the electrical impulse that controls the heart rate. However, these previous experiments could not fully reconstruct the tissue microenvironments, such as the propagation of an action potential (AP) or generation of contractions. To fulfill specific needs of heart-on-chip studies, a biohybrid construct was designed based on muscular thin films (MTFs); a tissue-engineered myocardium consist of anisotropic cardiomyocytes cultured on a deformable elastic thin film with various geometries [100, 101].

Grosberg et al. was pioneer in developing MTF-based "heart-on-a-chip" system that success-fully measured the contractility of neonatal rat ventricular cardiomyocytes exposed to various doses of epinephrine [100]. Eight separate MTFs were framed the skeleton of their system and was fabricated in batches enabling them to collect data from multiple tissues simultaneously in the same experiment. This heart-on-chip system mimicked the hierarchical tissue architecture of laminar cardiac muscle, and measurements of structure-function relationships, including contractility, AP propagation and cytoskeletal architecture. In another approach, Agarwal et al. explored an optimized semiautomated microdevice to test the positive inotropic effect of different dosages of isoproterenol on cardiac muscle contractility. They achieved an

increased drug-screening throughput with their device having 35 separate thin films
(**Figure 14a**) [102]. Basic components of this device includes a semiautomatic microdevice
integrated an MTF chip, an electrode for electric field simulation, a metallic base on a heating
element as temperature control unit and a transparent window for cantilever deformation
monitoring. As these models were based on animal tissues and cannot recapitulate human
system with precision. To overcome this limitation, Mathur et al. designed cardiac microphysiological system (MPS) that could imitate the human myocardium and envisage the cardio-
toxicity of drugs accurately, by merging hiPSC-derived (human-induced pluripotent stem cells
(hiPSC)) cardiomyocytes with an appropriate microarchitecture and "tissue-like" drug
gradients (**Figure 14b**) [103]. These hiPSC-derived cardiac MPS predicted drug response and
toxicity *in vitro* and showed a wide applicability for disease modeling and drug screening [82,
103]. Few reports are also available to tackle this complex yet vital organ of our system [104].

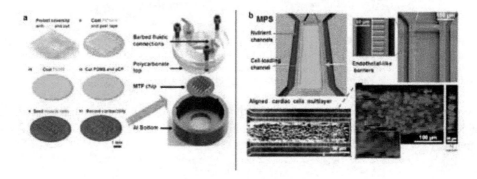

Figure 14. (a) Graphical illustration of the fabrication process flow for muscular thin film (MTF) and the semiautomatic
microdevice integrated a MTF chip [102]. (b) Schematic of the microphysiological system (MPS) with nutrient channels
(red), cell-loading channel (green) and 2 μm endothelial-like barriers. Optical and confocal fluorescence imaging of 3D
cardiac tissue aligned with multiple hiPSC cardiac cells layer [103].

3.2.5. Stem cells on chip

Human stem cells are a critical component for OOC devices. Few reports are available where
stems cells were grown in scaffolds and microarrays. These controlled conditions make it
possible to mimic the complex structures and cellular interactions within and between
different cell types and organs *in vivo* and keep the culture viable over long periods of time. It
was reported that neurogenesis of human mesenchymal stem cells can occur in the absence of
chemical stimuli, simply through the substrate stiffness [105]. **Figure 15a** is illustrating a PDMS
membrane-based platform for stem cells growth.

In principle, all cell sources, whether primary cells (directly taken from an organ or tissue, e.g.,
by means of a biopsy needle), or cells or in the form of cell lines, from animal or human origin,
can be useful for the OOC approach. The basic criterion for selecting the stem cells for OOC is
target disease. For the diseases with well-known gene mutation, the DNA, specific disease-
causing DNA mutation can be introduced into a stem cell line by the technique of homologous

recombination, resulting in two human cell lines with one having disease-causing DNA mutation in one of them. For this purpose, both hES (human stem cells) and iPSC (induced pluripotent stem cells) sources can in principle be used. On the other hand, for diseases caused by a whole spectrum of mutations in any part of the disease-causing gene, or diseases associated with a more complex genetic background, iPSC cell line or adult stem cells derived from a patient with the disease need to be used to recreate "the patient"—on a chip. iPSC cells are the first choice in contrast to adult stem cells, due to ease of regeneration [106].

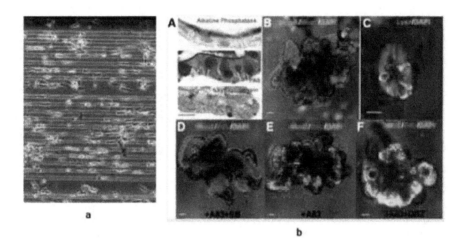

Figure 15. (a) PDMS based on-chip platform for stem cells. (b) Crypt-villus structures grown from single LGR5 positive adult stem cells from the intestinal crypt [106].

Three-dimensional "organoid" stem-cell culture technology was developed in the laboratory of Hans Clevers at the Hubrecht Institute. In this approach, intestinal stem cells were isolated from the intestinal epithelial tissue by separating tissue cells from each other. Subsequently, few stem cells within the cell mixture were identified by coupling them to a specific fluorescent antibody, followed by isolation with a fluorescence-activated cell sorter. 3D environment was created by the gel surrounding the cells to make them feel comfortable in their new "niche." In this process of cell growth, the stem cells were bound to their "mate," which is necessary to provide the essential cell-cell contact to start the self-renewal process. Once in the dish, each cell combination starts to self-assemble, a new crypt-villus structure in three dimensions forms called organoids (**Figure 15b**).

3.3. Anatomy-based organ function mimicking

As described in previous section, microengineering platforms evolved as critical methods for the fabrication of various models of organs in the biomedical sciences. Newer inventions in this filed are reported to generate patterns of complex microstructures with precise control of fluid dynamics and incorporation of specific biological element that simulates organ functions directly. In this segment, few OOC devices based on anatomical mimicking will be described.

3.3.1. Spleen on chip

Spleen is a secondary lymphoid organ for selective filtration of damaged RBCs and infectious microbes including *Plasmodium* parasites [107]. Keeping in mind its special role in filtration and to understand its functionally in deeper sense, it was critical to design an OOC with high precision and accuracy. Spleen consists of white pulp, red pulp, and the marginal zone and slow blood microcirculation through the reticular meshwork of the splenic red pulp with increasing hematocrit is the prime reason of its unique filtering capacity that facilitates specialized macrophages in recognizing and destroying unhealthy RBCs [108]. Rigat-Brugar-olas et al. designed a novel microdevice to copy the physical properties and hydrodynamic forces of the splenon; the minimal functional unit of the red pulp able to maintain filtering functions (**Figure 16a**) [108]. Their design consists of two main microfluidic channels for flow division to mimic the closed fast and the open slow microcirculations of spleen. The junction between slow-flow and fast-flow channel was arranged with parallel 2 μm microconstrictions resembling the IES to constrain cells. This device could precisely reproduce the natural physiochemical conditions of spleen and the unique characteristic of distinguishing different RBCs based on their mechanical properties.

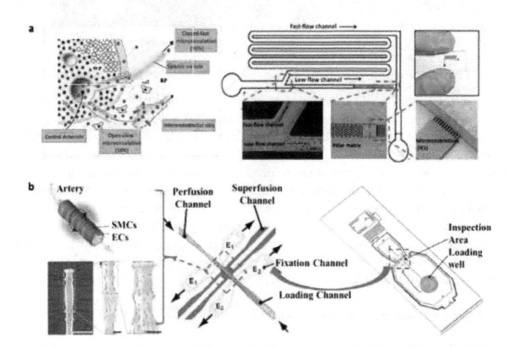

Figure 16. Splenon on a chip: (a-left) Diagram of the human splenon showing the closed-fast and open-slow microcir-culations as well as the interendothelial slits (IES); (a-right) Schematic representation of flow division zone, the pillar matrix and microchannels within slow-flow channel to mimic IES, respectively [108], (b) Artery on a chip: Schematic representation of a resistance artery segment on a chip contains a microchannel network, an artery loading well and an artery inspection area. ECs and SMCs represent the endothelial cells and smooth muscle cells, respectively [110].

3.3.2. Blood vessel on chip

Arteries or blood vessels transport the blood in human body. Geometry of vasculature and accumulation of particles inside the vessels varies with the pathological changes in the structure and function of small blood vessels, which leads to cardiovascular diseases [109]. Scalable approaches to assess the structure and function of intact cardiovascular tissues in health and disease will be crucial for developing better treatment strategies. Fluid sheer stress and cyclic stretch are other parameters that should be taken into account while designing *in vitro* vessels on-chip systems.

Typically, most of the current systems contain small arteries mounted on two wires or perfused with glass micropipettes that suffer from the disadvantages of nonscalability and need of a skilled person to operate. To overcome this barrier, Gunther et al. presented a scalable organ-based microfluidic platform for loading, precise placement, fixation as well as controlled perfusion and superfusion of a fragile resistance artery segment (**Figure 16b**) [110]. This device was comprised of three parts: the artery-loading area, a microchannel network and a separate artery inspection area, connected to a thermoelectric heater and a thermoresistor to maintain the temperature at 37°C. Resistance arteries had specialized structures with 30–300 μm diameters to regulate the flow and redistribution of blood in organs. As depicted, the setup was located in the terminal sections of the arterial vascular tree, and their walls are composed of a single layer of lining endothelial cells (ECs) and several layers of circumferentially arranged smooth muscle cells (SMCs). This device although could not replicate the full functionality but showed a unique property to analyze small artery structure and function through exposure to a well-defined heterogeneous spatiotemporal microenvironment.

In another approach by Zhang et al., cyclic stretching of vesicular endothelial cells can be studied. They designed a two-layered microsystem with upper microfluidics layer and bottom groove layer separated by an elastic membrane to provide cyclic stretch (**Figure 17**). A vacuum pump was integrated with the device to apply suction pressure on membrane resulting in cyclic stretch [111].

3.4. Perfusion-based on-chip systems

Cell-cell interactions are vital for maintaining tissue structure and function, and many cells respond to both homotypic and heterotypic interactions. Combining fluid flow and mechanical forcing regimens as in *in vivo* cellular environment can improve tissue- and organ-specific functions [66]. In this section, we describe few microengineering systems for liver, brain and womb that were designed for better understanding of mechanism of cellular interactions [82].

3.4.1. Liver on chip

Liver is considered to be one of the versatile organ performing thousands of functions that include detoxification, protein synthesis, hormone production, glycogen storage, etc. It is also a key player in human drug interaction and a trivial target for drug-induced toxicity.

Liver possess a complex structure and hepatic lobule is its prime functional unit consisting of hepatocytes, blood vessels, sinusoids and Kupffer cells. [112]. Hepatocytes are crucial con-

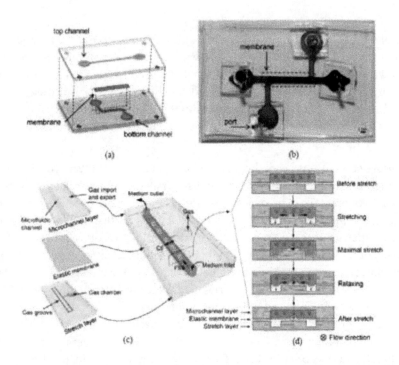

Figure 17. Schematic of blood vessel on chip. (a) PDMS chambers connected by a membrane, (b) Fabricated device, (c) Microfluidic channel for consecutive flow, (d) Stretching and relaxed elastic membrane [111].

tributors to liver functions and necessary for understanding the metabolism of xenobiotics and possible hepatotoxic effects in pharmacology. However, hepatocytes lack proliferative properties and biological interactions, which makes it rather difficult to maintain the liver-specific function of these cells *in vitro.* As a solution to this barrier, Kane et al. demonstrated a microfluidic array with wells capable of supporting micropatterned primary rat hepatocytes in coculture with 3T3-J2 fibroblasts [113]. In this process, under continuous perfusion with medium and oxygen, the synthetic and metabolic capacity of hepatocytes were preserved as evidenced by the continuous and steady synthesis of albumin and production of urea.

In other approach by Du et al., encapsulated hepatocytes that were produced with recombinant protein, with endothelial cells, differentiating them from hiPSCs within specific niches in multicomponent hydrogel fibers and further assembled into 3D-patterned endothelialized liver tissue constructs [114]. Endothelial cells significantly improved the function of hepatocytes *in vitro* and when tested on a mouse model of partial hepatectomy, an improved vascularization of the fiber scaffold was observed.

A miniaturized, multiwall coculture system for human hepatocytes surrounded by fibroblasts with optimized microscale architecture that maintained the typical phenotypic functions of the hepatocytes for several weeks was reported by Bhatia et al. Another device comprised of three sections, including a central channel for heptocytes, a microfluidics convection channels and a microfluidics sinusoid barrier with a set of narrow channels to model epithelial cells as show in **Figure 18a** [115]. This model succeeded in mimicking the transportation between blood flow and hepatocytes and the sheer stress experienced by hepatocyts.

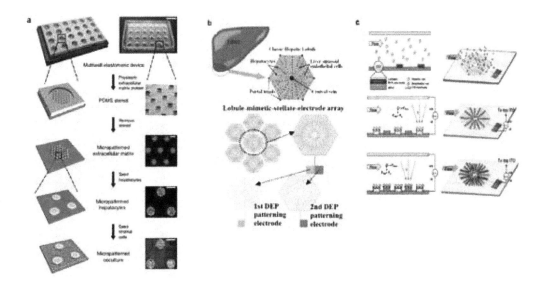

Figure 18. (a) Schematic of soft lithographic process to fabricate microscale multiwell format for primary hepatocytes that selectively adhere to matrix-coated domains and coculture with fibroblasts seeded on bare areas [115]. (b) Configuration of one basic unit of liver tissue, the classic hepatic lobule and lobule-mimetic-stellate-electrodes array. (c) The configuration and operation principles of DEP-based heterogeneous lobule-mimetic cell patterning [116].

Another research, also based on hepatocytes-based model, was done by Ho et al., where they designed an array of concentric-stellate-tip microelectrodes to mimic the lobular structure of liver tissues (**Figure 18b, c**) [116]. This device was comprised of vertical microelectrodes or lobule-mimicking stellate electrode arrays, to achieve 3D liver cell patterning by separately snaring hepatocytes and endothelial cells that were manipulated under patterned electric fields via dielectrophoresis (DEP). Few other researchers (i.e., Feng et al. [117], Wong et al. [118], Lee et al. [119]) have also put forth their proof of concepts based on hepatocytes. Wong et al. developed a concave microwell-based size controllable spheroidal "hepatosphere" and "heterosphere" models by monoculturing primary hepatocytes and by coculturing primary hepatocytes and hepatic stellate cells (HSCs), respectively, to monitor the effect of HSCs in controlling the formation of tight cell-cell contacts and final organization of the spheroidal aggregates [118, 82].

Some other reports are also there where researcher came forward with their ideas to design efficient liver on-chip devices for drug screening and toxicity analysis [98, 120–122].

Recently, Lee et al. have designed a novel liver on-chip system based on liver microsomes that were encapsulated in 3D hydrogel matrix to mimic the metabolism reactions and the transport phenomena in the liver. Photopolymerization of poly(ethylene glycol) diacrylate (PEG-DA) allowed controlling the mass transfer with matrix sizes. To reproduce the blood flow through liver, gravity-induced passive flow was explored. They measured the reaction kinetics of P450 enzymes in the device and simulated the convection-diffusion-reaction characteristics inside the device with a mathematical model [123]. **Figure 19a** is illustrating the schematic and design of on-chip liver platform. Although there were several factors to be modified for improved reaction kinetic data such as diffusion limitation, optimization of convection and mixing, reducing the nonspecific binding to PDMS surface, preliminary analysis shows great potential and this device will be further explored for the metabolism of various compounds in liver [123].

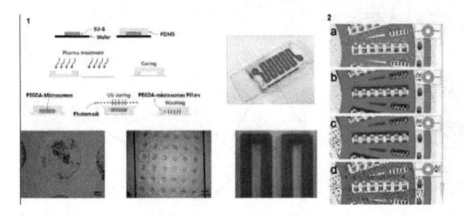

Figure 19. (1) Schematic of PDMS chip fabrication method and picture of fabricated chip (size of the glass slide was 25 mm by 75 mm) [123]. (2) Schematic diagram illustrating the sequential procedure for constructing the biomimetic microtissue [124].

Most of the on-chip liver platforms are based on hepatocytes, and generally, these *in vitro* hepatocyte culture systems imitated the structure of the hepatic cord or can applied for studying specific aspects of toxicity. However, to imitate advanced liver architectures (i.e., hepatic sinusoids) that could preserve cell-cell and cell-ECM interactions, these existing devices did not solve the purpose. To overcome this limitation, Ma et al designed a microfluidics-based biomimetic method for *in vitro* fabrication of a 3D liver lobule such as microtissue. Their system was composed of a radially patterned hepatic cord-like network and an intrinsic hepatic sinusoid-like network as shown in **Figure 19b**. This device showed that the 3D biomimetic liver lobule-like microtissue retained higher basal liver-specific functions in Phase I/II (i.e., CYP-1A1/2 and UGT activities) and more sensitive response was obtained for pharmacological inducers/inhibitors than the 2D and 3D monocultures of HepG2 cells. This device was tested for three model drugs—acetaminophen, isoniazid and rifampicin and a high hepatic capacity for drug metabolism was exhibited by biomimetic microtissue that indicated that microtissue, designed by Ma et al. can be explored as a promising platform for *in vitro* toxicity of drugs [124].

3.4.2. Brain on chip

Human brain is the most complex structure and the quest to understand how it stores and processes information leads researches to the application of new microengineering technologies to design *in vitro* model of brain. Unraveling the basic concepts could be beneficial for neural diseases, development of improved brain-machine interfaces and domain of machine-learning will be totally revolutionized. A brain-oriented paradigm shift has occurred with the advances in neuroscience and OOC systems [131, 132].

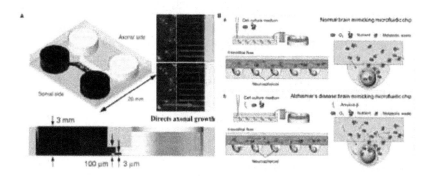

Figure 20. (A) The microfluidic-based culture platform directs axonal growth of CNS neurons and fluidically isolates axons [125]. (B) Schematic diagrams of normal brain mimicking microfluidic chip (a) and Alzheimer's disease brain mimicking microfluidic chip [127].

We discussed earlier various other OOC but owing to its structural and functional hierarchy, high specialization and constant metabolic demand to design a complete *in vitro* brain model is difficult. The prime limiting factors are used to identify the smallest structural and functional unit, ion channels or synapses in the microenvironment [126]. Researchers from all over the world give different experimental models of circular microfluidic compartmentalized coculture platforms to study brain development and degeneration based on physiological neuron connection architecture. A microfluidic culture platform was demonstrated by Taylor et al, consist of a relief pattern of somal and axonal compartments connected by microgrooves that function in directing, isolation and biochemical analysis of CNS axons (**Figure 20a**) [125]. In another work, Park et al describe a microfluidic chip based on 3D neurospheroids that more closely mimics the *in vivo* brain microenvironment and provides a constant flow of fluid similar in the interstitial space of the brain. Concave microwell arrays were explored for the formation of uniform neurospheroids, with cell-cell interactions and contacts in all directions while osmotic micropump was used to maintain the slow interstitial level of flow. Using this platform, effect of flow on neurospheroid size, neural network and neural differentiation was investigated via this *in vitro* platform. Larger sizes of neurospheroids were obtained and formed more robust and complex neural networks than those cultured under static conditions. This finding proved the effect of the interstitial level of slow and diffusion-dominant flow on continuous nutrient, oxygen and cytokine transport and removal of metabolic wastes [127]. This chip was designed to detect the toxic effect of β-amyloid; a major contributor of Alzheimer's disease. **Figure 20b** is showing the schematic of this unique platform for neurodegenerative disease diagnostic. Kato-Negishi et al. came up with a millimeter-sized neural building block to reconstruct 3D broad neural networks connecting with different neurons [128]. Peyrin et al. also described a microfluidic system involving several different neuron subtypes separated into two individual chambers with asymmetrical connection architecture of funnel-shaped microchannels to reconstruct oriented neuronal networks [129]. This device was a kind of diode that operated as direction selective filter where axonal projections can be penetrated by axons in a single direction and as an impermeable barrier for cell bodies. In this point, Kunze et al. demonstrated a 3D microfluidic device for creating physiologically realistic, micrometer scaled neural cell multilayers in an alginate-enriched agarose scaffold [82, 130].

A method to fabricate neurospheres networking with nerve-like structure using concave well arrays connected by the hemicylindrical channels was illustrated by Jeong et al. This method provides the topological effect of the concave-well hemicylindrical-channel-networking, which is crucial in guided outgrowth of neuronal network [131]. Similar hemicylindrical systems were also explored to generate 3D nerve-like neural bundles between neural spheroids and neighboring satellite spheroids in concave channels [132].

3.4.3. Breast and womb on chip

Breast cancer is still the cause of concern and with the advancement of microfabrication techniques, improved detection and therapy of breast neoplasia can be obtained via nanodevices traveling inside mammary ducts. However, the decreasing size of branched mammary ducts prevents access to remote areas of the ductal system using a pressure-driven fluid-based approach. Magnetic field guidance of superparamagnetic submicron particles (SMPs) in a stationary fluid might provide a possible alternative but it is critical to first reproduce the breast ductal system to assess the use of such devices for future therapeutic and diagnostic ("theranostic") purposes. Graften et al. came up with an idea of to engineer a portion of a breast ductal system using polydimethylsiloxane (PDMS) microfluidic channels of decreasing sizes with a total volume of 0.09 mL. A magnet was used to move superparamagnetic/fluorescent SMPs through a static fluid inside the microchannels [133]. **Figure 21a** is the schematic of PDMs on chip assembly. This device can be explored for the early detection of ductal breast cancer and consisted of basoapically polarized monolayer of luminal cells only as the device imitated the luminal portion of the ductal breast system only and myoepithelial cells at the basal side of the luminal cells and terminal ductal lobular units at the ends of the narrowest channels were not included. Apart from breast on chip, womb OOC was also developed by Chang et al. with the objective to deal with infertility.

In recent years, a genuine increase in infertility has been observed due to diverse factors, including stress, environmental pollution and increase in age, smoking, consumption of alcohol, sexually transmitted diseases, etc. *In vitro* fertilization (IVF), a state-of-the-art technology, enhances the rate of pregnancy. As a procedure, fertilized eggs in the blastocyst stage are transferred to the woman's uterus for implantation and further development and efforts are made to improve the culture environment of the preimplantation embryos and developing specialized culture surfaces to enhance the success rate of this technique [134, 135].

Due to the failure of static culture systems to mimic the dynamic fluid environment in the fallopian tube [136], dynamic culture platforms that explored shaking/rotation [137], controlled fluid flow [138] and vibration [139] models were studied for use in embryonic development where method of coculturing embryos with endometrial was done to overcome developmental arrest of early embryos in single culture. Although these methods showed enhanced perform-

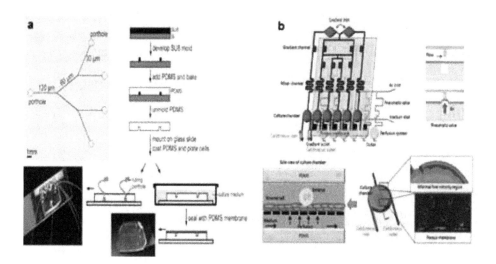

Figure 21. (a) Schematic of breast on chip [133], (b) PDMS-based embryo coculture microchip, where the concentration gradient generator is integrated with a mixer and a cell culture chamber on the top [140].

ance and beneficial effects of coculturing on the development of mammalian, they could not be considered as a complete on chip system for womb. Recently, Chang et al. also designed an autologous 3D perfusion platform as a necessary approach to deal with IVF and partly mimic the physiological function of the reproductive system [140]. This device as shown in **Figure 21b** is comprised of an upstream concentration gradient generator (width: 250 μm, height: 230 μm) was integrated with a diamond-shaped passive micromixer (width: 200 μm, height: 230 μm) that could generate six different homogeneous concentrations of progesterone. Micromixer was used to increase the contact area between liquid molecules and to provide enhanced mixing efficiency by its continuous splitting and mixing of liquids. The main specifications and goals of this microfluidic channel design was as follows: (i) Gradient distribution for specific concentrations of steroid hormones in six culture chambers, (ii) Maintaining homogenous concentrations of steroid hormones in individual chamber, (iii) Preserve uniform culture conditions with respect to the flow speed/rate by constant flow speed/rate for the chambers.

This womb-on-chip platform showed the ability to replace the present embryo culture platforms used for assisting *in vitro* fertilization.

3.5. Human on chip

Organ-on-chip concept is in its nascent state and despite of the substantial advances in the creation of microengineered tissue and organ models, a lot is left to explore for recreating complex 3D models that could reconstitute the whole organ metabolism and physiology. With the recent advances in tissue engineering, microfabrication techniques, researchers are now focusing on multiorgan-on-chip devices that could imitate complete human on chip up to some extent if not fully [141, 142]. Figure shows a body-on-chip systems.

Although complete functional body-on-chip devices are still far from reach but the latest development in this field has given a glimpse of promising future of this revolutionary field of biomedicine. **Figure 22** shows the concept of body-on-chip microsystem [4].

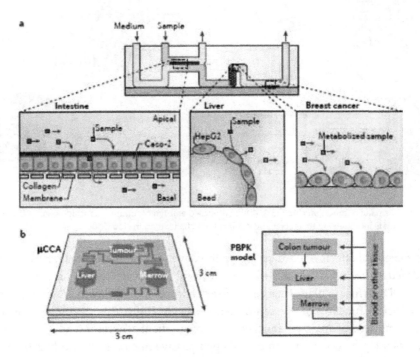

Figure 22. Schematic of a body on chip system. (a) A microdevice containing interconnected cell culture microchambers integrated with microfluidic culture of intestinal epithelial, hepatocytes and breast cancer cells. (b) A micro cell culture analog (μCCA) representing a colon tumor, the bone marrow and liver [4].

4. Pharmaceutical applications and future prospects of organ-on-chip devices

The field of OOC devices is still in its infancy, although it is a rapidly growing research arena with lots of future potential in biomedicine from understanding the mechanism of complex organ architectures to drug discovery. Earlier studies revealed that while 3D cell cultures were far more superior planar than conventional 2D models due to their better control over cell differentiation, ECM mechanical compliance and a much better response was obtained in terms of tissue- and organ-level functionality by combining microengineering with cell biology. Fortunately, with the recent advances in microfabrication strategies and microfluidics, precise dynamic control of structure, mechanics and chemical delivery at the cellular size scale can be achieved. Microengineered 3D cell culture models, and particularly more sophisticated OOC microdevices, have many potential applications, including disease research and drug discovery, but in this section, we mainly focus on OOC application relevant to pharmaceutical industry.

The pharmaceutical industry is under intense pressure economically, ethically and scientifically to find ways to accelerate the drug-development process, and to develop drugs that are safer and more effective in humans at a lower cost. Traditional animal testing approaches are expensive and often fail to predict human toxicity or efficacy of drugs; in fact, nowadays, questions are arising with regard to the significance of animals testing if they cannot reliably predict clinical outcomes [4, 143]. As correctly suggested by Dr. Ingber, Founder Director-Harvard's Wyss Institute for Biologically Inspired Engineering that chips respond to drugs like human organs do — —and have the potential to replace animal testing for safety and efficacy early in the drug-development process.

4.1. Bottlenecks in drug discovery process

4.1.1. High cost of compound testing

Modern drug development requires implementation of extensive preclinical testing and validation protocols before getting the formal approval to progress to clinical evaluation of the compound. This process is tedious and costly and a single compound can cost more than $2 million. Moreover, every 10 drugs entering clinical trials, generally only one or two would be licensed for eventual use in humans [2].

4.1.2. Lack of exact simulation of human systems in static 2D cells culture

The lack of preclinical model systems to provide accurate predictions of human responses to novel therapeutic drugs is another critical limiting factor in drug discovery. The current gold standard for laboratory-based preclinical evaluation is based on *in vitro* cell culture assay and *in vivo* animal model experimentation and assessment. Although cell culture assays have advantage of controlled environments where cellular maturation and activity are easily observed and tested, they lack the complexity of living systems and are incapable of mimicking the conditions of organ-organ or tissue-tissue communication. This simplicity is a major drawback in drug-development studies since drug metabolism and the effect of metabolite activity on nontarget tissues cannot be predicted [3].

4.1.3. Time period of animal studies and loss of numerous animal lives

Another crucial limiting factor is time involved in *in vivo* studies. Although animal studies can somehow better predict the drug metabolism and response as animal models maintain the intricacy of living systems and assessment of organ-organ crosstalk and nontarget organ toxicity is possible , these models on multiple occasions, been proven to be wrong predictors of human responses to drug treatment. Human system is more complex and developed than laboratory animals and the response and mechanisms are different for many therapeutic agents. The hypothesis that favorable outcomes observed in animals will translate to human patients has led to clinical situations where treatments have proved futile or even detrimental to patient well-being and recovery [3, 144].

4.1.4. Lack of accurate prediction of clinical response and diminished number of new drugs for patients

As discussed earlier, due to inadequate *in vitro* data and practical difficulties of *in vivo* studies, the clinical response is not always as expected. Eroom's law (Moore's law backwards) states that "the number of new medicines *halves* every nine years," despite an "astronomical" increase in research funding from government and industry. This situation exists in large part because the traditional journey from drug discovery to drug development still occurs mostly in 2D static cell cultures and animal studies, which are not the true predictors of response of new compounds in the human body resulting in failure of approximately 85% of therapies in clinical trials and of those that make it to advanced phase III, generally the last step before regulatory approval, only half are actually approved. This data itself ignite are concerns for the pharma industry and how to expedite the current drug discovery scenario [149].

Microengineered cell culture systems that mimic complex organ physiology have the potential to be used for the development of *in vitro* human-relevant disease models. These are more predictive of drug efficacy and toxicity in patients and can provide better insight into drug mechanism of action. OOC devices provide compelling advantages over other *in vitro* cell culture models for the evaluation of drug safety and metabolism. In broader sense, *in vitro* assays incorporating cultured human cells can act as savior in identifying environmental toxins and providing better understanding of their mechanisms of action, as well as improving our ability to predict risks for specific compounds. In addition, the ability to integrate functional organ mimetics, such as gut, liver, lung and skin-on-chips within a "human-on-a-chip," the interplay of different organs in determining pharmacokinetic properties of compounds can be monitored [3, 145].

4.2. Role of organ-on-chip devices in drug discovery

4.2.1. Reduction in cost

The drug-development process is costly in the phases of clinical trials, which can cost millions of dollars. However, despite extensive animal testing of drugs before starting a clinical trial with humans, many drugs fail because of low efficacy or unexpected toxic side effects not predicted with earlier trials. In this regard, the most promising advantage of body-on-a-chip devices is that the devices can mimic both animal and human metabolism and predict differences between them that will allow for a higher level of accuracy when predicting the outcome of clinical trials. Moreover, any toxicity observed before human trial with *in vitro* on chip systems can prevent unsuitable drug candidates from entering the expensive phase of clinical trials that limit costs and unrealistic expectations.

Body-on-a-chip devices are low-cost platforms that can substantially reduce the cost of drug testing.

4.2.2. Drug-target identification

Organs-on-chips have the potential to serve as a new enabling platform to identify and validate the effectiveness, safety of potential targets early in the pipeline to increase the likelihood of success in clinical trials [4]. Song et al has recently a microengineered model of vasculature to mechanistically examine chemokine-mediated interactions between circulating breast cancer cells and the microvascular endothelium that induced site-specific basal stimulations and activation of the microfluidic endothelium by introducing chemokines into the lower chambers. Through quantitative analysis of cancer cell attachment to the endothelium and the levels of cell surface receptor expression, this system predicted that endothelial recruitment of breast cancer cells induced by a chemokine-CXC-chemokine ligand 12 (CXCL12), involved in cancer metastasis, is mediated by the endothelial receptor CXCR4 and this response is independent of the expression of CXCL12 receptors on circulating cancer cells. These findings gave a new insight into critical role of the vascular endothelium in the metastatic behavior of circulating tumor cells and how to control and manipulate a biological target to analyze a functional outcome of target modulation. This discovery related with OOC model was an important breakthrough in indentifying a valid therapeutic target for preventing cancer metastasis [146].

Other studies on OOC platforms for understanding of molecular mechanisms of cell-cell interactions, mitochondrial cardiomyopathy of Barth syndrome, and drug-induced toxicities in pulmonary edema have also been successfully performed [147–149].

4.2.3. Toxicity and drug efficacy evaluation

This a very important aspect of drug research as toxicity analysis is utmost important for any new therapeutic agent. Liver and kidney tissues are of great interest to drug developers due to their predominant role during the absorption, distribution, metabolism and excretion (ADME) process of a drug [3]. Physiologically, drug is metabolized mainly in the liver while kidney deals with their elimination. These two critical processes make these two organs highly susceptible to drug injury. In a coculture bio-analytical microplatform of liver-kidney, toxicity of anticancer drug ifosfamide illustrated the importance of the liver-kidney interaction. Ifosfamide is a prodrug, activated in body system by CYP450 enzymes in the liver, but some of its metabolites, such as chloracetaldehyde, are nephrotoxic. With this model of highly differentiated liver cells (HepaRG), perturbation of cell proliferation and calcium release in the kidney tissue could be monitored that was not possible with the single culture. Previously, the same group simulated the performance of hepatocytes on-chip system coupled with NMR for toxicity analysis of flutamide [149, 150].

These contributions signify the role of on-chip systems for toxicity analysis of drug *in vitro* that is an important step for clinical trials.

Multiorgan interactions in drug testing and their importance were highlighted by Sung et al. also. They studied the dose response and efficacy of 5-fluorouracil (5-FU) on a system containing system that contained liver cells (HepG2/C3A), colon cancer cells (HCT-116) and myeloblasts (Kasumi-1) [151]. They monitored the degradation phenomenon of 5-Fu and effect of its pro drug Tegafur and uracil-a competitive inhebitor of 5-Fu for the dose response and bioavailability.

Predicting the bioavailability of a drug accurately can be difficult with animal models. Multiorgan microdevices that contain a combination of the gastrointestinal tract epithelium and the liver at the appropriate sizes and with realistic liquid-to-cell ratios have the potential to predict the bioavailability of ingested drugs [152].

4.2.4. Drug screening

The absence of predicted therapeutic effects of a drug or increased dose levels is the major cause of drug toxicity. The failure of existing methods to accurately predict *in vivo* drug efficacy before clinical trials give rise to the undesirable outcomes. Human OOC models can become instrumental in addressing these existing imitations [4].

The potential of OOC approaches for testing drug efficacy was recently explored by Aref et al. in a microengineered 3D assay of epithelial-mesenchymal transition (EMT) during cancer progression [153]. By culturing lung cancer spheroids in a 3D matrix gel adjacent to an endothelialized microchannel, this model recapitulated EMT-induced tumor dispersion and phenotypic changes in cancer cells in an endothelial cell-dependent manner. Twelve drugs ranging from prospective drugs to US Food and Drug Administration (FDA)-approved drugs were screened into the vascular channel, and their ability to inhibit EMT was analyzed by direct visualization of the cancer spheroids.

The results obtained for drugs efficacy in cancer treatment by on-chip systems, significantly varies from 2D static culture and were in close proximity with human clinical trials. This study concluded that such OOC systems will be developed as a more realistic platform for efficacy and to decide for advanced trails, a major step toward drug discovery.

4.2.5. Response of combination of drugs

Since microdevices are relatively inexpensive, and many such devices will be operated in parallel, it is possible to test many drugs and combinations of drugs at different concentrations with devices. Testing combinations of drugs is useful to monitor drug interactions and cross talks. Synergistic interactions are of particular interest. Another benefit of such studies is that the drugs having similar functions, but different side effects could potentially be combined at reduced dosages to achieve the needed tissue response. These multiorgan on-chip systems can play a major role to design individualized therapy regimen for patients that do not respond to routinely used drug combinations as a synergistic effect and dose of different drug combination can be predicted.

4.2.6. Pharmacokinetics and body on-chip systems

Physiologically based pharmacokinetic models (PBPKs) are mathematical models that are used to extrapolate data from animal experiments and predict human response to a drug. These models mainly rely on existing understanding and knowledge of a drug's metabolism from traditional 2D static cultures and animal studies and as we discussed, these methods are not the accurate predictors. This is the reason for the equations used in a PBPK are not complete and the models are not accurate. Multiorgan microdevices can be modeled more precisely with PBPKs and divergence between the model's prediction and experimental data obtained with the devices can enhance our understanding of human response to a wide variety of combination of inputs with higher accuracy than before.

To generate a precise PBPK model, for pharmacokinetics and pharmacodynamics studies, recapitulating human physiology at the whole-body level is the most crucial aspect. Researchers have begun to pursue the development of multi-organ models, and in one such study, combined models of breast cancer, the intestine and liver were designed to create a network of interconnected microfabricated cell culture chambers that exhibited the sequential absorption, metabolism and efficacy of four anticancer drugs [154]. Shuler et al [155] applied pharmacokinetic and pharmacodynamic modeling (PKPD) principles to micro cell culture analog comprising interconnected microchambers representing a colon tumor, the liver and bone marrow, which imitated the *in vivo* distribution, retention and recirculation of drug-containing blood in these organs. Hepatic metabolism-mediated cytotoxicity of the prodrug tegafur to colon cancer, liver cancer and bone marrow cells was investigated by this system. These multiorgan on-chip systems are better than the existing models and can expedite the drug discovery process by increasing the efficiency and mitigating the high cost associated with drug-development process.

4.3. Future prospects of organ-on-chip devices

As an alternative to conventional cell culture and animal models, human OOC could transform many areas of basic research and drug development. They have wide applications in research on molecular mechanisms of organ development and disease, organ-organ coupling and the interactions of the body with stimuli, such as drugs, environmental agents, consumer, products and medical devices. Due to complexities involved, OOC have limited or no applications in certain areas of biomedical research, such as chronic diseases, adaptive immune responses or complex system-level behaviors of the endocrine, skeletal and nervous systems. As described previously, OOC are effective for investigating physiological and disease processes that occur in a relatively short-time frame (less than ~1 month) and depend on relative cell positions within an organ- or tissue-specific microarchitecture [66].

OOC technology has certain technical and entrepreneurial challenges also. One of the critical technical challenges is material for fabrication—such as poly(dimethylsiloxane) (PDMS) that have gained widespread use in rapid-prototyping of OOC microdevices as most of the OOC models rely mostly on synthetic materials (e.g. PDMS, polycarbonate and polyester), the physicochemical properties of which are not appropriate for mimicking extracellular matrices *in vivo*. It is utmost important to identify new cell culture substrates to produce devices for more accurate predictions. For successful translation of OOC from proof of concept in the laboratory to commercial screening platforms, identification and optimization of new low-cost materials and fabrication strategies suitable for their mass production and integration into existing infrastructures in the pharmaceutical industry is call of time.

More reliable and sustainable sources of human cells, especially disease-specific cells that are acquiescent to *in vitro* culture in OOC and phenotypically are true representative of their *in vivo* counterparts are required. To overcome this hurdle, human embryonic stem cells and iPS cells can be engineered to suit specific needs in the development of OOC [3, 156]. The OOC models with stem cells can generate and control physiologically relevant structural, biochemical and mechanical cues required for stem-cell differentiation and maturation.

With the new avenues opened by OOC in drug development, there is a need of fabricating human on-chip or multiorgan on-chip devices and to maintain a balance between the complexity and practicality will play an important role in their wide applications. With the improvement in physiological relevance, complexity in the model is obvious that presents major challenges to practical operation and management of the system. Accurate identification of minimal subset of cells and microenvironmental factors will be helpful to create a balance and designing a simplest model possible that recapitulates physiological responses of interest.

Integration of laboratory on-chip platforms with miniaturized analytical systems is also important for better detection sensitivity despite of low culture volumes and cell numbers [1].

OOCs are not universal solutions, and alternative tools will continue to be better solutions for modeling certain *in vivo* processes as animal offer whole-organism toxicity testing and this parallel analysis will be required until the current OOC scenario attains the maturity and refine human on-chip systems come into existence.

Despite their limitations, OOCs have the potential to play a transformative role across drug discovery and development. Eventually, OOC models may play a pivotal role in streamlining the clinical trial process. Due to the complexities of organ function and regulatory requirements, it is unlikely that OOCs will replace animal testing anytime soon [66].

However, with the scientific advancements, this field is evolving at a fast pace and these hurdles could be surmountable with tri-lateral partnerships between academic institutions, industry and regulatory agencies. The paradigm-shifting potential of OOC technology has been recognized by funding agencies integrated microphysiological systems [157, 158]. Pharmaceutical companies are also coming forward to establish industry- academia partnerships to jointly explore this emerging research arena and to establish themselves at the forefront of expected OOC advances. In nut shell, it is concluded that despite of several limitations, achievements in this revolutionary field of biomedicine, OOC technology present exciting new avenues for drug discovery and development and a perfect picture of a promising future.

Acknowledgements

Author is grateful to Department of Science and Technology, Government of India for providing INSPIRE Faculty award to pursue independent research.

Author details

Preeti Nigam Joshi

Address all correspondence to: ph.joshi@ncl.res.in

Organic Chemistry Division, National Chemical Laboratory, Pune, India

References

[1] Neužil P, Giselbrecht S, Länge K, Huang TJ, Manz A. Nat Rev Drug Discov. 2012; 11: 620–632. DOI: 10.1038/nrd3799

[2] Dimasi JA, Grabowski HG. The cost of biopharmaceutical R&D: is biotech different? Manage Decis Econ. 2007; 28: 469–479. DOI: 10.1002/mde.1360

[3] Esch MB, Smith A, Prot JM, Sancho CO, Hickman J, ShulerML. Adv Drug Deliv Rev. 2014; 0: 158–169. DOI: 10.1016/j.addr.2013.12.003

[4] Esch EW, Bahinski A, Huh D. Nat Rev Drug Discov. 2015; 14: 248–260. DOI:10.1038/nrd4539

[5] Shamir ER, Ewald AJ. Nat Rev Mol Cell Biol. 2014; 15: 647–664. DOI: org/10.1016/j.tcb.2015.09.010

[6] Whitesides GM, Ostuni E, Takayama S, Jiang XY, Ingber DE. Annu Rev Biomed Eng. 2001; 3: 335–373. DOI: 10.1146/annurev.bioeng.3.1.335

[7] Craighead H. Nature. 2006; 442: 387–393. DOI:10.1038/nature05061

[8] Dhaliwa A. Mater Method. 2012; 2: 162. DOI: http://dx.DOIorg/10.13070/mm.en.2.162

[9] Yang SI, Lei KF, Tsai SW, Hsu HT. Conf Proc IEEE Eng Med Biol Soc. 2013; 2013: 168–171. DOI: 10.1109/EMBC.2013.6609464

[10] Patolsky F, Zheng G, Lieber CM. Nat Protoc. 2006; 1: 1711–1724. DOI: 10.1038/nprot.2006.227

[11] Zhang C, Xing D. Nucleic Acids Res. 2007; 35: 4223–4237. DOI: 10.1093/nar/gkm389

[12] Ashraf MW, Tayyaba S, Afzulpurkar N. J Mol Sci. 2011; 12: 3648–3704. DOI: 10.3390/ijms12063648.

[13] Wu J, Kodzius R, Cao W, Wen W. Microchim acta. 2013; 81: 1611–1616. DOI: 10.1007/s00604-013-1140-2

[14] Daar AS, Thorsteinsdóttir H, Martin DK, Smith AC, Nast S, Singer PA. Nat Genet. 2002; 32: 229–232. DOI: 10.1038/ng1002-229

[15] Yang J, Liu Y, Rauch CB, Stevens RL, Liu RH, Lenigk R, Grodzinski P. Lab Chip. 2002; 2: 179–187. DOI: 10.1039/B208405H

[16] Zhou X, Liu D, Zhong R, Dai Z, Wu D, Wang H, Du Y, Xia Z, Zhang L, Mei X, Lin B. Electrophoresis. 2004; 25: 3032–3032. DOI: 10.1002/elps.200305966

[17] Cady NC, Stelick S, Kunnavakkam MV, Batt CA. Sens Actuat B Chem. 2005; 107: 332–341. DOI: 10.1016/j.snb.2004.10.022

[18] Wu J, Kodzius R, Xiao K, Qin J, Wen W. Biomed Microdevice. 2012; 14: 179–186. DOI: 10.1007/s10544-011-9595-6

[19] Chia BT, Yang SA, Cheng MA, Lin CL. IEEE 22nd International Conference. 2009; 292–295. DOI: 10.1109/MEMSYS.2009.4805376

[20] Steinbach C, Steinbrücker C, Pollok S, Walther K, Clement JH, Chen Y, Petersen I, Cialla-May D, Weber K, Popp J. Analyst. 2015; 140: 2747–2754. DOI: 10.1039/C4AN02086C

[21] Kopp MU, Mello AJ, Manz A. Science. 1998; 15: 1046–1053.

[22] Yeo LY, Chang HC, Chan PPY, Friend JR. Small. 2011; 7: 12–48. DOI: 10.1002/smll.201000946

[23] Ahn CH, Choi JW, Beaucage G, Nevin JH, Lee JB, Puntambekar A, Lee JY. Proc IEEE. 2004; 92:154–173.

[24] Service RF. Science. 1998; 282: 399–401.

[25] Dinh T, Griffin G, Stokes D, Wintenberg A. Sens Actuators B. 2003; 90: 104–111. DOI: 10.1016/S0925-4005(03)00048-0

[26] Kricka L. Clin Chim Acta. 2001; 307: 219–223. DOI:10.1016/S0009-8981(01)00451-X

[27] Lauks I. Acc Chem Res. 1998; 31: 317–324. DOI: 10.1021/ar9700670

[28] Ko J. Lab Chip. 2003; 3: 106–113. DOI: 10.1039/B301794J

[29] Linder V. Analyst. 2007; 132: 1186–1192. DOI: 10.1039/B706347D

[30] Stevens W, Gous N, Forf N, Scott LE. BMC Medicine2014; 12:173–180. DOI: 10.1186/s12916-014-0173-7

[31] Choi S, Goryl M, Sin LYM, Wong PK, Chae J. Microfluid Nanofluid 2011; 10: 231–247. DOI: 10.1007/s10404-010-0638-8

[32] Su W, Gao X, Jiang L, Qin J. J Chromatograph A. 2015; 1377: 13–26. DOI:10.1016/j.chroma.2014.12.041

[33] Khan IU, Serra CA, Anton N, Vandamme T. J Control Release. 2013; 172: 1065–1074. DOI: 10.1016/j.jconrel.2013.07.028

[34] Khan IU, Serra CA, Anton N, Vandamme T. Expert Opin Drug Deliv. 2015; 2: 547–562. DOI: 10.1517/17425247.2015.974547

[35] Stadler B. Biomicrofluidics. 2015; 9: 052501. DOI: 10.1063/1.4931070

[36] Hettiarachchi K, Zhang S, Feingold S, Lee AP, Dayton PA. Biotechnol Prog 2009; 25: 938–945. DOI: 10.1002/btpr.214

[37] Wu Y, Li L, Mao Y, Lee LJ, ACS Nano. 2012; 6: 2245–2252. DOI: 10.1021/nn204300s

[38] Jahn A, Vreeland WN, Gaitan M, Locascio LE. J Am Chem Soc 2004; 126: 2674–2675. DOI: 10.1021/ja0318030

[39] Kim S, Kim HJ, Jeon NL. Integr Biol. 2010; 2: 584–603. DOI: 10.1039/c0ib00055h

[40] Abhyankar VV, Lokuta MA, Huttenlocher A, Beebe DJ. Lab Chip. 2006; 6: 389–393. DOI: 10.1039/b514133h

[41] Kricka LJ, Fortina P. Clin Chem. 2001; 47: 1479–1482.

[42] Wang L, Li PCH. Anal Chim Acta. 2011; 687: 12–27. DOI: 10.1016/j.aca.2010.11.056

[43] Liu RH, Dill K, Fuji HS, McShea A. Expert Rev Mol Diagn. 2006; 6: 253–261.

[44] Sosnowski RG, Tu E, Butler WF, OConnell JP, Heller MJ. Proc Natl Acad Sci USA. 1997; 94: 1119–1123.

[45] Lee HH, Smoot J, McMurray Z, Stahl DA, Yager P. Lab Chip. 2006; 6: 1163–1170. DOI: 10.1039/B605507A

[46] Zheng B, Tice JD, Roach LS, Ismagilov RF. Angew Chem Int Ed Engl. 2004; 43: 2508–2511. DOI: 10.1002/anie.200453974

[47] Zhang B, Tice JDm, Ismagilov RF. Anal Chem. 2004; 76: 4977–4982. DOI: 10.1021/ac0495743

[48] Huang S, Li C, Lin B, Qin J. Lab Chip. 2010; 10: 2925–2293. DOI: 10.1039/c005227b

[49] Yager P, Edwards T, Fu E, Helton K, Nelson K, Tam MR, Weigl BH. Nature. 2006; 442. DOI: 10.1038/nature05064

[50] Harrison RG, Greenman MJ, Mall FP, Jackson CM. Anat Rec. 2005; 1: 116–128. DOI: 10.1002/ar.1090010503

[51] Bhise N, Gray RC, Sunshine JC, Htet S, Ewald AJ, Green JJ. Biomaterials. 2010; 31: 8088–8096. DOI: 10.1016/j.biomaterials.2010.07.023

[52] Wang Z, Samanipour R, Koo K, Kim K. Sens Mater. 2015; 27: 487–506.

[53] Breslin S, O'Driscol L. Drug Discov Today. 2013; 18: 240–249. DOI:10.1016/j.drudis.2012.10.003

[54] Huh D, Hamilton GA, Ingber DE. Trends Cell Biol. 2011; 21: 745–754. DOI:10.1016/j.tcb.
      2011.09.005

[55] VaheriA, Enzerink A, Räsänen K, Salmenperä P. Exp Cell Res. 2009; 315: 1633–1638.
      DOI:10.1016/j.yexcr.2009.03.005

[56] Eiraku M, Watanabe K, Matsuo-Takasaki M, Kawada M, Yonemura S, Matsumura M,
      Wataya T, Nishiyama A, Muguruma K, Sasai Y. Cell Stem Cell. 2008; 3: 519–532. DOI:
      10.1016/j.stem.2008.09.002

[57] Sutherland RM. J Natl Cancer Inst. 1971; 46: 113–120. DOI: 10.1093/jnci/46.1

[58] Yuhas JM, Li AP, Martinez AO, Ladman AJ. Cancer Res. 1977; 37: 3639–3643.

[59] Matsuda N, Shimizu T, Yamato M, Okano T. Adv Mater. 2007; 19: 3089–3099. DOI:
      10.1002/adma.200701978

[60] Kawaguchi N, Hatta K, Nakanishi T. Biomed Res Int. 2013; 2013: 895967. DOI:
      10.1155/2013/895967

[61] Slamon DJ, Godolphin W, Jones LA, Holt JA, Wong SG, Keith DE, Levin WJ, Stuart SG,
      Udove J, Ullrich A. Science. 1989; 244: 707–712.

[62] Choi SH, Kim YH, Hebisch M, Sliwinski C, Lee S, D'Avanzo C, Chen H, Hooli B, Asselin
      C, Muffat J, Klee JB, Zhang C, Wainger BJ, Peitz M, Kovacs DM, Woolf CJ, Wagner SL,
      Tanzi RE, Kim DY. Nature. 2014; 515: 274–281. DOI: 10.1038/nature13800

[63] Pickl M, Ries CH. Oncogene. 2009; 28: 461–468. DOI: 10.1038/onc.2008.394

[64] Tung YC, Hsiao AY, Allen SG, Torisawa YS, Ho M, Takayama S. Analyst. 2011; 136:
      473–478. DOI: 10.1039/c0an00609b

[65] Debnath J, Brugge JS. Nat Rev Cancer. 2005; 5: 675–688. DOI:10.1038/nrc1695

[66] Bhatia SN, Ingber DE. Nat Biotechnol. 2014; 32: 760–772. DOI: 10.1038/nbt.2989.

[67] Kim HJ, Huh D, Hamilton G, Ingber DE. Lab Chip. 2012; 12: 2165–2174. DOI: 10.1039/
      C2LC40074J

[68] Kim HJ, Ingber DE. Gut-on-a-Chip microenvironment induces human intestinal cells
      to undergo villus differentsiation. Integr Biol (Camb). 2013; 5: 1130–1140. DOI: 10.1039/
      c3ib40126j

[69] Huh D, Matthews BD, Mammoto A, Montoya-Zavala M, Hsin HY, Ingber DE. Science.
      2010; 328: 1662–1668. DOI: 10.1126/science.1188302

[70] Jang K, Suh KY. Lab Chip. 2010; 10: 36–42. DOI: 10.1039/B907515A

[71] Ma Z, Wang J, Loskill P, Huebsch N, Koo S, Svedlund FL, Marks NC, Hua EW,
      Grigoropoulos CP, Conklin BR, Healy KE. Nat Commun. 2015; 14: 7413. DOI: 10.1038/
      ncomms8413

[72] Sticker D, Rothbauer M, Lechner S, Hehenberger MT, Ertl P. Lab Chip. 2015 Dec 21; 15(24): 4542–4554. DOI: 10.1039/c5lc01028d

[73] Lee PJ, Hung PJ, Lee LP. Biotechnol Bioeng. 2007; 97: 1340–1346. DOI: 10.1002/bit.21360

[74] Esch MP, Sung JH, Yang J, Yu C, Yu J, March JC, Shuler ML. Biomed Microdevices. 2012; 14: 895–906. DOI: 10.1007/s10544-012-9669-0

[75] You L,Temiyasathit S, Lee P, Kim CH, Tummala P, Yao W, Kingery W, Malone AM, Kwon RY, Jacobs CR. Bone. 2008; 42: 172–179. DOI: 10.1016/j.bone.2007.09.047

[76] Sudo R, Chung S, Zervantonakis IK, Vickerman V, Toshimitsu Y, Griffith LG, Kamm RD. FASEB J. 2009; 23: 2155–2164. DOI: 10.1096/fj.08-122820

[77] Park J, Koito H, Li J, Han A. Biomed Microdevices. 2009; 11: 1145–1153. DOI: 10.1007/s10544-009-9331-7

[78] Halldorsson S, Lucumi E, Gómez-Sjöberg R, Fleming RMT. Biosens Bioelectron. 2015; 63: 218–231. DOI: 10.1016/j.bios.2014.07.029

[79] Jiang S, Cheng R, Wang X, Xue T, Liu Y, Nel A, Huang Y, Duan X. Nat Comm. 2013; 2225. DOI: 10.1038/ncomms3225

[80] Pereira-Rodrigues N, Poleni P-E, Guimard D, Arakawa Y, Sakai Y, Fujii T. PLoS ONE. 2010; 5: e9667. DOI: 10.1371/journal.pone.0009667

[81] Xu H, Ferreira MM, Heilshorn SC. Lab Chip. 2014; 14: 2047–2056. DOI: 10.1039/c4lc00162a

[82] Zeng F, Fu F, Cheng Y, Wang C, Zao Y, Gu Z. Small. 2016. DOI: 10.1002/smll.201503208

[83] Verpoorte E, Oomen P, Skolimowski M, Mulder P, van Midwoud P, Starokozhko V, Merema M, Molema G, Groothuis G. IEEE. 2015; 224–227.

[84] Yum K, Hong SG, Healy KE, Lee LP. Biotechnol J. 2014; 9: 16–27. DOI: 10.1002/biot.201300187

[85] Czupalla CJ, Liebner S, Devraj K. Methods Mol Biol. 2014; 1135: 415–437. DOI: 10.1007/978-1-4939-0320-7_34

[86] Hatherell K, Couraud PO, Romero IA, Weksler B, Pilkington GJ. J Neurosci Methods. 2011; 199: 223–229. DOI: 10.1016/j.jneumeth.2011.05.012

[87] Jang KJ, Cho HS, Bae WG, Kwon TH, Suh KY. Integr Biol. 2011; 3: 134–141. DOI: 10.1039/C0IB00018C

[88] Huh D, Fujioka H, Tung TH, Futai N, Paine R, Grotberg JB, Takayama S. Proc Natl Acad Sci USA. 2007; 104: 18886–18891. DOI: 10.1073/pnas.0610868104

[89] Tavana H, Zamankhan P, Christensen PJ, Grotberg JB, Takayama S. Biomed Microdevices. 2011; 13; 31–742. DOI: 10.1007/s10544-011-9543-5

[90] Douville NJ, Zamankhan P, Tung YC, Li R, Vaughan BL, Tai CF, White J, Christensen P, Grotberg JB, Takayama S. Lab Chip. 2011; 11: 609–619. DOI: 10.1039/C0LC00251H

[91] Schetz M, Dasta J, Goldstein S, Golper T. Curr Opin Crit Care. 2005; 11: 555–565.

[92] Weinberg E, Kaazempur-Mofrad M, Borenstein J. Int J Artif Organs. 2008; 31: 508–514.

[93] Jang KJ, Suh KY. Lab Chip. 2010; 10: 36–42. DOI: 10.1039/b907515a

[94] Jang KJ, Mehr AP, Hamilton GA, McPartlin LA, Chung S, Suh KY, Ingber DE. Integr Biol (Camb). 2013; 5: 1119–1129. DOI: 10.1039/C3IB40049B

[95] Benson K, Cramer S, Galla H. CNS. 2013; 10: 5. DOI: 10.1186/2045-8118-10-5

[96] Ma SH, Lepak LA, Hussain RJ, Shain W, Shuler ML. Lab Chip. 2005; 5: 74–85. DOI: 10.1039/B405713A

[97] Shayan G, Choi YS, Shusta EV, Shuler ML, Lee KH. Eur J Pharm Sci. 2011; 42: 148. DOI: 10.1016/j.ejps.2010.11.005

[98] Brown JA, Pensabene V, Markov DA, Allwardt V, Neely MD, Shi M, Britt CM, Hoilett OS, Yang Q, Brewer BM. Biomicrofluidics. 2015; 9: 054124. DOI: 10.1063/1.4934713

[99] Booth R, Kim H. Lab Chip. 2012; 12: 1784–1792. DOI: 10.1039/c2lc40094d

[100] Agarwal A, Farouz Y, Nesmith AP, Deravi LF, McCain ML, Parker KK. Adv Funct Mater. 2013; 23: 3738–3746. DOI: 10.1002/adfm.201203319

[101] Grosberg A, Nesmith AP, Goss JA, Brigham MD, McCain ML, Parker KK. J Pharmacol Toxicol Methods. 2012; 65: 126–135. DOI: 10.1016/j.vascn.2012.04.001.

[102] Agarwal A, Goss JA, Cho A, McCain ML, Parker KK. Lab Chip. 2013; 13: 3599–3608. DOI: 10.1039/C3LC50350J

[103] Mathur A, Loskill P, Shao K, Huebsch N, Hong S, Marcus SG, Marks N, Mandegar M, Conklin BR, Lee LP. Sci Rep. 2015; 5: 8883. DOI: 10.1038/srep08883

[104] Jastrzebska E, Tomecka E, Jesion I. Biosens Bioelectron. 2016; 75: 67–81. DOI: org/10.1016/j.bios.2015.08.012

[105] Wang LS, Chung JE, Chan PPY, Kurisawa M. Biomaterials. 2010; 31: 1148–1157. DOI: 10.1016/j.biomaterials.2009.10.042

[106] Organ on chip meeting report. Lab Chip. 2013; DOI: 10.1039/c3lc50248a

[107] Herricks T, Seydel KB, Molyneux M, Taylor T, Rathod PK. Cell Microbiol. 2012; 14: 1880–1891. DOI: 10.1111/cmi.12007

[108] Rigat-Brugarolas L, Elizalde-Torrent A, Bernabeu M, Niz MD, Jaular LM, Fernandez-Becerra C, Homs-Corbera A, Samitier J, del Portillo H. Lab Chip. 2014; 14: 1715–1724. DOI: 10.1039/C3LC51449H

[109] Maron BJ, Wolfson JK, Epstein SE, Roberts WC. J Am Coll Cardiol. 1986; 8: 545–557. DOI:10.1016/S0735-1097(86)80181-4

[110] Günther A, Yasotharan S, Vagaon A, Lochovsky C, Pinto S, Yang J, Lau C, Bolz JV, Bolz SS. Lab Chip. 2010; 10: 2341–2349. DOI: 10.1039/C004675B

[111] Zheng W, Ziang B, Wang D, Zhang W, Wang Z, X Ziang. Lab Chip. 2012; 12: 344–3450. DOI: 10.1039/c2lc40173h

[112] Geerts A. Semin Liver Dis. 2001; 21; 311–335. DOI: 10.1055/s-2001-17550

[113] Kane BJ, Zinner MJ, Yarmush ML, Toner M. Anal Chem. 2006; 78: 4291–4298. DOI: 10.1021/ac051856v

[114] Du C, Narayanan K, Leong MF, Wan AC. Biomaterials. 2014; 35: 6006–6014. DOI: 10.1016/j.biomaterials.2014.04.011

[115] Khetani SR, Bhatia SN. Nat Biotechnol. 2008; 26: 120–126. DOI: 10.1038/nbt1361

[116] Ho CT, Lin EZ, Chen RJ, Chin CK, Gong SE, Chang HY, Peng HL, Hsu L, Yew TR, Chang SF. Lab Chip. 2013; 13: 3578–3588. DOI: 10.1039/c3lc50402f

[117] Feng ZQ, Chu XH, Huang NP, Leach MK, Wang G, Wang YC, Ding YT, Gu ZZ. Biomaterials. 2010; 31: 3604–3612. DOI: 10.1016/j.biomaterials.2010.01.080

[118] Wong SF, Choi YY, Kim DS, Chung BG, Lee SH. Biomaterials. 2011; 32: 8087–8096. DOI: 10.1016/j.biomaterials.2011.07.028

[119] Lee SA, No DY, Kang E, Ju J, Kim DS, SH Lee. Lab Chip. 2013; 13: 3529–3537. DOI: 10.1039/C3LC50197C

[120] Goral VN, Hsieh YC, Petzold ON, Clark JS, Yuen PK, Faris RA. Lab Chip. 2010; 10: 3380–3386. DOI: 10.1039/C0LC00135J

[121] Prodanov L, Jindal R, Bale SS, Hegde M, McCarty WJ, Golberg I, Bhushan A, Yarmush ML, Usta OB. Biotechnol Bioeng. 2015; 113: 241–246. DOI: 10.1002/bit.25700

[122] Domansky K, Inman W, Serdy J, Dash A, Lim MH, Griffith JG. Lab Chip. 2010; 10: 51–58. DOI: 10.1039/B913221J

[123] Lee J, Kim SH, Kim YC, Choi I, Sung JH. Enzyme Microb Technol. 2013; 53: 159–164. DOI: 10.1016/j.enzmictec.2013.02.015

[124] Ma C, Zhao L, Zhou E, Xu J, Shen S, Wang J. Anal Chem. 2016; 88: 1719-1727. DOI: 10.1021/acs.analchem.5b03869

[125] Taylor AM, Blurton-Jones M, Rhee SW, Cribbs DH, Cotman CW, Jeon NL. Nat Methods. 2005; 2: 599–605.

[126] Millet LJ, Gillette MU. Trends Neurosci. 2012; 35: 752–776. DOI: 10.1016/j.tins.2012.09.001

[127]   Park J, Lee BK, Jeong GS, Hyun JK, Leeae CJ, Lee SH. Lab Chip. 2015; 15: 141–150. DOI: 10.1039/C4LC00962B

[128]   Peyrin JM, Deleglise B, Saias L, Vignes M, Gougis P, Magnifico S, Betuing S, Pietri M, Caboche J, Vanhoutte P. Lab Chip. 2011; 11: 3663–3673. DOI: 10.1039/C1LC20014C

[129]   Negishi MK, Morimoto Y, Onoe H, Takeuchi S. Adv Healthc Mater. 2013; 2: 1564–1570. DOI: 10.1002/adhm.201370058

[130]   Kunze A, Giugliano M, Valero A, Renaud P. Biomaterials. 2011; 32: 2088–2098. DOI: 10.1016/j.biomaterials.2010.11.047

[131]   Jeong GN. Tissue Eng Regen Med. 2014; 11: 297–303. DOI: 10.1007/s13770-014-4047-z

[132]   Jeong GS, Chang JY, Park JS, Lee SA, Park D, Woo J, An H, Lee CJ, Lee SH. Mol Brain. 2015; 8: 17. DOI: 10.1186/s13041-015-0109-y

[133]   Grafton MMG, Wang L, Vidi PA, Leary J, Lelie'vre SA. Integr Biol. 2011; 3: 451–459. DOI: 10.1039/c0ib00132e

[134]   Swain LE, Smith GD. Hum Reprod Updat. 2011; 17: 541–557. DOI: 10.1093/humupd/dmr006

[135]   Smith GD, Rocha AM. Semin Reprod Med. 2012; 30: 214–221. DOI: 10.1055/s-0032-1311523

[136]   Quinn P. Culture Media, Solutions, and Systems in Human ART, CambridgeUniversity Press, India, 2014

[137]   Isachenko V, Montag M, Isachenko E, van der Ven K, Dorn C, Roesing B. Reprod Biomed Online. 2006; 13: 228–234.

[138]   Fukui Y, Lee ES, Araki N. J Anim Sci. 1996; 74: 2752–2758. DOI: /1996.74112752x

[139]   Hur Y, Park J, Ryu E, Park S, Lee J, Lee S. J Assist Reprod Gen. 2013; 30: 835–84. DOI: 10.1007/s10815-013-0007-0

[140]   Chang KW, Chang PY, Huang HY, Li CJ, Tien CH, Yao DJ, Fan SK, Hsu W, Liu CH. Sensor Actuator B. 2016; 226: 218–226. DOI: org/10.1016/j.snb.2015.11.004

[141]   Maschmeyer I, Lorenz AK, Schimek K, Hasenberg T, Ramme AP, Hübner J, Lindner M, Drewell C, Bauer S, Thomas A, Sambo NS, Sonntag F, Laustera R, Marxab U. Lab Chip. 2015; 15: 2688–2699. DOI: 10.1039/C5LC00392J

[142]   Maschmeyer I, Lorenz A, Ramme A, Hasenberg T, Schimek K, Hubner J, Lauster R, Marx U. Abstr-Toxicol Lett. 2015; 238S: S56–S383. DOI: 10.1016/j.toxlet.2015.08.277

[143]   Scannell JW, Blanckley A, Boldon H, Warrington B. Nat Rev Drug Discov. 2012; 11: 191–200. DOI: 10.1038/nrd3681

[144]   Greek R, Menache A. Int J Med Sci. 2013; 1: 206–221. DOI: 10.7150/ijms.5529

[145] Esch MB, Mahler GJ, Stokol T, Shuler ML. Lab Chip. 2014; 14: 3081–3092. DOI: 10.1039/C4LC00371C

[146] Song JW, Cavnar SP, Walker AC, Luker KE, Gupta M, Tung YC, Luker GD, Takayama S. PLoS ONE. 2009; 4: e5756. DOI: org/10.1371/journal.pone.0005756

[147] Kunze A, Lengacher S, Dirren E, Aebischer P, Magistretti PJ, Renaud P. Integr Biol (Camb). 2013; 5: 964–975. DOI: 10.1039/c3ib40022k

[148] Wang G, et al. Nat Med. 2014; 20: 616–623. DOI: 10.1038/nm.3545

[149] Huh D, Leslie DC, Matthews BD, Fraser JP, Jurek S, Hamilton GA, Thorneloe KS, McAlexander MA, Ingber DE. Sci Transl Med. 2012; 4: 159ra147. DOI: 10.1126/scitranslmed.3004249

[150] Choucha-Snouber L, Aninat C, Grsicom L, Madalinski G, Brochot C, Poleni PE, et al. Biotechnol Bioengin. 2012; 110: 597–608. DOI: 10.1002/bit.24707

[151] Sung JL, Kam C, Shuler ML. Lab Chip. 2010; 10: 446-455. DOI: 10.1039/B917763A

[152] Snouber LC, Bunescu A, Naudot M, Legallais C, Brochot C, Dumas ME, Herrmann BE, Leclerc E. Toxicol Sci. 2013; 132: 8–20. DOI:10.1093/toxsci/kfs230

[153] Aref AR, Huang RY, Yu W, Chua KN, Sun W, Tu TY, Bai J, Sim WJ, Zervantonakis IK, Thiery JP, Kamm RD. Integr Biol (Camb). 2013; 5: 381–389. DOI: 10.1039/c2ib20209c

[154] Imura Y, Sato K, Yoshimura E. Anal Chem. 2010; 82: 9983–9988. DOI: 10.1021/ac100806x

[155] Sung JH, Shuler ML. Lab Chip. 2009; 9: 1385–1394. DOI: 10.1039/B901377F

[156] Mathur A, Loskill P, Hong S, Lee J, Marcus SG, Dumont L, Conklin BR, Willenbring H, Lee LP, Healy KE. Stem Cell Res Ther. 2013; 4 Suppl 1: S14. DOI: 10.1186/scrt375

[157] Sutherland ML, Fabre KM, Tagle DA. Stem Cell Res Ther. 2013; 4 (Suppl. 1): I1. DOI: 10.1186/scrt361

[158] Bernard A, Reanult JP, Michel HR, Bosshard HR, Delamarche E. Adv Mat. 2000; 12: 1067. DOI: 10.1002/1521-4095(200007)12:14<1067::AID-ADMA1067>3.0.CO;2-M

[159] Therriault D, White SR, Lewis JA. Nat Mat. 2; 2003: 265–271. DOI:10.1038/nmat863

4

Fabrication of Three-Dimensional Concave or Convex Shell Structures with Shell Elements at Micrometer Resolution in SU-8

Louis WY Liu, Qingfeng Zhang and Yifan Chen

Abstract

This chapter presents a photo-lithographically-based technology for mass production of three-dimensional (3D) micro-structures with shell elements. In this technology, shell elements are photo-lithographically fabricated at micron or sub-micron resolution by illuminating with ultraviolet light radiating an ultraviolet light beam onto UV-opaque SU-8 monomer. The technology does not require any steps involving micro-injection molding or micro-stereolithography. Several prototypes have been fabricated to demonstrate the feasibility of this technology.

Keywords: SU-8, Micro-fabrication, Injection molding, Micro-stereolithography, Graytone lithography

1. Introduction

Micrometer features with hollow parts are increasingly common in many applications [1]. Micro-needles, drug delivery systems, and vacuum micro-electronics operating at millimeter-wave frequencies all have hollow micro-structures with dimensions that extend from half millimeter to 1 μm scale. However, mass producing hollow objects at micron resolution is a known challenge in the field of micro-fabrication [2]. Preliminary information about the enabling technologies for realization of suspended or hollow micro-structures is discussed in the sections which follow.

2. Micro-injection molding

Micro-injection molding is currently the technology most widely used for mass-fabrication of three-dimensional (3D) hollow micro-structures [3,4,11–13]. In micro-injection molding, thermoplastic granules are first melted in the plastifying unit of a micro-injection machine. Then, the molten plastic is injected at high pressure into the hollow space of an injection molding tool. After a cooling process, the injection tool is disassembled and the molded object removed. Although micro-injection molding is an established technology that supports mass production of hollow objects, the process is not without limitations. To date, the sizes of hollow objects made in conventional micro-injection molding technology are typically in millimeter range, not in micron range. Injection molding of an object at millimeter resolution is not possible without a special injection machine and auxiliary equipment. The mold to be used for formation of object at millimeter resolution has to be equipped with inlets and outlets in order to allow high-speed injection, gas evacuation, and the expulsion. More importantly, the process of micro-injection molding involves many energy intensive steps which are eco-unfriendly.

3. Micro-stereolithography

Another most widely used 3D micro-fabrication technology is micro-stereolithography [1,2,8]. It works by scanning an UV laser on a liquid monomer, curing the monomer into solid polymeric slices layer by layer, and stacking together all these polymeric slices with various contours. This UV-induced photo-polymerization repeats in a layer-by-layer fashion until the desired 3D object is fully formed. This technology has made it possible to fabricate any form of 3D micro-structures. The surface profile of a fabricated micro-structure can be as complicated as a human face. However, micro-stereolithography is a time-intensive process. The typical scanning speed of a micro-stereolithography machine is about 200–300 layers per hour [8], depending on the geometry and the resolution of the 2D slice to be formed on each layer. Fabricating a simple 3D object of 1 mm in height can take more than 30 minutes. Fabricating a small array of micro-needles can take anywhere between 50 minutes and several hours. Micro-stereolithography technology is currently being pushed developed aggressively focusing on for improvements in both resolutions, speed and flexibility in choice of photo-curable materials. However, due to the use of a laser and a scanner system, the initial investment costs of a micro-stereolithography-based process are unavoidably high.

4. Graytone lithography

Graytone lithography [9,10] is another inexpensive 3D micro-fabrication technology. This technology has been evolved to one step mask-less fabrication using SU-89. For the applications not requiring or not suited with no access to using an expensive micro-stereolithography machine, gray tone lithography is an interesting alternative to micro-stereolithography. Graytone lithography is a modification of conventional 2D optical lithography. It works by exposing a positive photoresist to a UV light through a grayscale mask which defines the patterns of 3D micro-structures to be formed. The UV light through the grayscale mask produces local intensity modulation. Following a UV exposure, the 3D profile on the surface of the positive photoresist can be formed on the substrate by stripping off the UV-exposed photoresist. While this technology is well known for its potential for mass production, the technology itself is not without limitation. One such limitation is that grayscale lithography based on a positive photoresist does not support fabrication of hollow or suspended micro-structures without change in the technology.

5. Comparison between different 3D micro-fabrication technologies

To the authors' knowledge, to date, there has not been any cost-effective approach dedicated to mass production of hollow micro-structures at micron or sub-micron resolution. In this paper, the proposed fabrication process presented is a fundamentally different approach based on an improved version of another process published in references [5,6]. The proposed process can be carried out photo-lithographically with conventional photo-lithographic equipment. It provides a convenient alternative for researchers without no access to a micro-stereolithography machine or other expensive fabrication facilities. It is not intended as a replacement of other already established and accessible 3D micro-fabrication technologies. Instead, we believe that the proposed process should be used in conjunction with other 3D micro-fabrication technologies to optimizing maximize the advantages in speed, precision, repeatability, and costs of manufacture. Unlike micro-stereolithography, which is only suitable for fabricating small objects in the micron range, the proposed technology can be used to fabricate larger objects with dimension in excess of 1 mm with no sacrifice of speed. The fabricated structures or micro-structures are optically smooth. **Table 1** summarizes the advantages and disadvantages in comparison with other competing technologies.

	Micro-stereolithography [1,2,8]	Micro-injection molding [11–13]	Grayscale lithography [9,10]	This technology [5–7]
Initial investment	High	High	Low	Low
Materials	A negative photocurable material	A thermoplastic	A positive photo-curable material	A UV-opaque SU-8
Processing speed	Depends on the thickness of the structure	Fast	Fast	Fast
Feasibility for fabricating shell or hollow components	Yes	Yes	Not easy	Yes

	Micro-stereolithography [1,2,8]	Micro-injection molding [11–13]	Grayscale lithography [9,10]	This technology [5–7]
Suitability for fabricating large objects	No	Yes	Yes	Yes
Resolution	Micron range	Millimeter range	Micron range	Micron range
Suitability for mass-production	No	Yes	Yes	Yes

Table 1. Comparison between different 3D micro-fabrication technologies.

6. SU-8 combined with a UV-blocking impurity

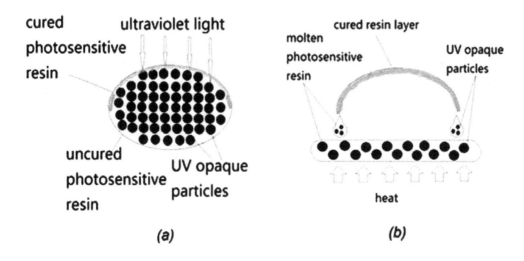

Figure 1. (a) A cross-sectional view showing that SU-8 mixed with a UV-opaque impurity is exposed to ultraviolet light. The gray region represents the UV-exposed area where cross-linking of SU-8 monomers occurs and (b) a cross-sectional view showing that the UV-exposed structure from **Figure 1a** is subjected to heat. The uncured SU-8 resin melts into a liquid because of the elevated temperature.

Standard SU-8 fabrication processes normally require a UV-transparent SU-8 [14–17] to proceed. To start fabricating a micro-structure with shells or suspended layers, however, we need a custom-made SU-8 monomer. This SU-8/impurity composite essentially contains a mixture of an SU-8 monomer and a UV-opaque impurity. The UV-opaque impurity is an organic chemical compound satisfying the following criteria: i) it must be opaque to UV lights, ii) it must be highly soluble in SU-8, and iii) it must be substantially non-adhesive. Our experimental results suggest store-bought plasticines or synthetic rubber Blu-Tack™ can be used as an example of the UV-opaque impurities in the proposed process. In this study, Blu-Tack plasticine is the UV-opaque impurity added to the SU-8 to increase its UV opacity. The ratio of SU-8 to Blu-Tack is 5:3 by volume. The mixture is then baked at 110°C for 4 hours to evaporate off the majority of the solvent (i.e. GBL).

Before exposure to any UV light, it is essential to ensure that the SU-8/impurity composite can be reshaped upon heating and retain its chemical properties after cooling down. However, when this SU-8/impurity composite is exposed to a UV light, as shown in **Figure 1a**, only the surface exposed on which the ultraviolet light has shone to can be polymerized into in a thin layer. The thickness of this layer can be changed by changing the duration of the UV exposure. In general, the longer the UV exposure, the thicker this layer will be.

Since the UV light cannot reach the SU-8/impurity composite underneath the UV exposed surface, the SU-8/impurity composite underneath the UV-exposed surface will remain uncured. The heat then applied to this experimental setup will melt this uncured SU-8/impurity composite into a liquid, as illustrated in **Figure 1b**. Since the removed SU-8/impurity composite removed remains photo-curable, it can be reused to fabricate other hollow parts.

7. Fabrication of an embossing stamp

Another key element for mass production of micro-structures is an embossing stamp, as shown in **Figure 2e**. The embossing stamp serves as a patterned template for casting the desired 3D patterns on the surface of the SU-8/impurity composite. A variety of methods can be used to fabricate this embossing stamp. If the 3D features to be formed have many complicated contours, then gray tone lithography or micro-stereolithography [1–4] should be used. If the 3D patterns to be formed are just a combination of several arbitrarily profiled sidewalls, then the method illustrated in [5,7,18] may be adopted.

8. Casting of three-dimensional micro-structures and formation of suspended or hollow parts

Once the embossing stamp and the SU-8/impurity are available, we can proceed to replicate the desired 3D micro-structures and fabricate suspended and hollow parts. This process is illustrated in **Figure 2a–h**.

In step 1, as shown in **Figure 2a**, several height-defining blocks of equal height are fabricated onto the corners of the wafer. These height-defining blocks are used to ensure the SU-8 monomer to be deposited on the wafer becomes even and accurate in thickness.

In step 2, as shown in **Figure 2b**, the SU-8/impurity composite is deposited onto the surface of the substrate until its thickness slightly exceeds the height of the height-defining blocks. The wafer is then heated so that its temperature is slightly above about 70–80°C. The SU-8/impurity composite is heated at this temperature for a prolonged period until the SU-8/impurity composite is void of any solvent.

In step 3, as shown in **Figure 2c**, the top surface of the SU-8/impurity is pressurized and ironed flat with a glass slice. Then, the heat source is removed.

In step 4, as shown in **Figure 2d**, patterns defining the hollow or suspended regions are metalized. This metallization process can be carried out by painting with metal ink followed by etching. The purpose of this step is to create an embedded mask which allows selective UV exposure in the later step. After the UV exposure step, the embedded mask created in this step will become a sacrificial layer to be removed in the final step.

In step 5, as shown in **Figure 2e**, the embossing stamp fabricated in the previous stage is manually aligned with a mask-aligner and pressed downwards slowly. This step not only casts the conical pattern on the right of **Figure 2e**, but it also forms hemispherical solids by pressing the metal layers downwards. Since the top surface of the semi-molten SU-8/impurity composite has its own surface tension, this step will ensure that a smooth surface with a spherical profile is formed on the top of the hemispherical package.

In step 6, as shown in **Figure 2f**, the embossing stamp is removed from the wafer. The wafer is then cooled down for about 3 hours until the 3D patterns on the SU-8/impurity composite become fully solidified. This step further eliminates the surface adhesion on the top of the wafer.

In step 7, as shown in **Figure 2g**, areas requiring hollow or suspended parts are selectively exposed to a UV light. This step can be realized by two methods. We can expose the wafer to an ordinary UV light through a photomask that defines the patterns of hollow or suspended parts. Since the SU-8/impurity composite has been mixed with a UV-opaque impurity, the UV-exposed surface of the SU-8/impurity composite can be polymerized into a polymeric layer. This polymeric layer can be easily thickened by increasing the duration of the UV exposure. This polymeric layer can be as thin as a membrane or as thick as the application demands. The SU-8/impurity underneath the UV-exposed surface will remain uncured and become removable by melting.

In step 8, as shown in **Figure 2h**, the wafer is baked at 110°C. This heat temperature not only hardens the polymerized surface from step 7 but also melts the uncured SU-8/impurity mixture underneath the UV-exposed surface, into a liquid. Traces of the SU-8/impurity not removable by melting can be further removed by developing in an appropriate SU-8 developer.

The fabricated component on the top left corner of **Figure 2h** is a hemispherical package used to protect RF-MEMS devices against humidity. This hemispherical package has been glued onto a metallic washer which serves as a radiofrequency ground for a radiofrequency-printed circuit board. The package is intended to be capped on the substrate of a printed circuit board on which the RF-MEMS devices are mounted. The interface between the hemispherical package and the substrate can be further sealed with PDMS.

The fabricated component on the top middle **Figure 2h** is a spherical hollow object with a diameter equal to approximately 500 μm. It is realized by gluing together two hemispherical shell elements fabricated using the same process as illustrated in **Figure 2a–h**. Gluing the two hemispherical shell elements together involves manual alignment under a microscope.

The component in the top right corner of **Figure 2h** is a conical funnel with base diameter equal

to approximately 50 μm.

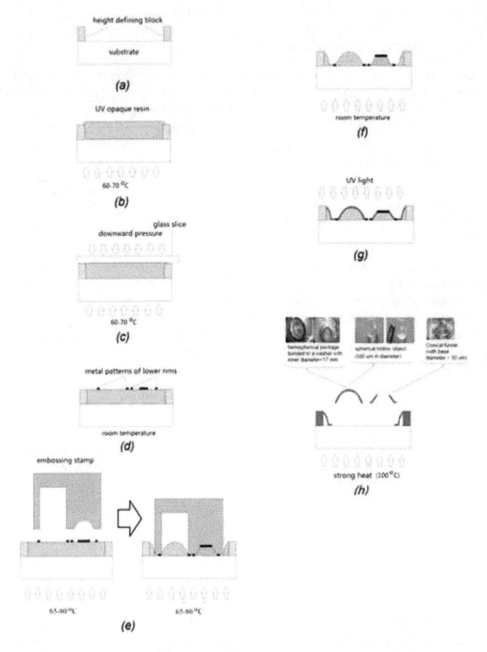

Figure 2. (a–h) Cross-sectional view illustrating the process flow of the technology for fabricating hollow or suspended micro-structures.

9. Fabricating devices for drug delivery applications

Over the past decades, there has been no shortage of interest in nano- or micro-fabrication in the field of drug delivery. Among all drug delivery devices, micro-needles appear to be the most popular apparatus in terms of the number of papers that have been published over the past three decades. With the above fabrication technique, fabricating a micro-needle or similar devices would be as easy as brewing coffee. Formation of hollow objects at micro-scale is undoubtedly one of known challenges in the field of micro-fabrication, especially polymer micro-fabrication. The following sections demonstrate how the previously developed techniques [14–19] can be employed for realization of conical micro-funnels for drug delivery applications. As explained in the previous sections, our fabrication methodology is based purely on photo-lithography.

In step 1, as shown in **Figure 3a**, we need to prepare a pre-cured SU-8 which is UV opaque. A standard SU-8 monomer resin is first mixed with a UV-opaque impurity at an elevated temperature until the final mixture becomes almost opaque in the UV spectrum. This UV-opaque impurity can be an appropriate plasticine that is opaque to UV lights and does not form any complex in SU-8. In addition, this UV-opaque SU-8 resin will have to undergo a prolonged dehydration bake to increase its viscosity and to decrease its surface adhesion. The wafer of this pre-cured SU-8 will be maintained at an elevated temperature (preferably slightly above the glass transition temperature) so that this UV-opaque SU-8 resin becomes partially molten and its upper surface becomes non-adhesive. In so doing, this UV-opaque SU-8 resin should be able to be reshaped upon heating without excessive change in its physical and chemical properties.

In step 2, as shown in **Figure 3b**, we need to fabricate an embossing stamp which is basically a master mold having an array of cylindrical rods. This embossing stamp will be used as a patterned template for casting of the inverted conical patterns on the surface of the UV-opaque SU-8 from step 1. A variety of methods can be employed to fabricate this embossing stamp. In the present study, an embossing stamp with high aspect ratio micro-rods was fabricated using a high-quality standard UV-lithographic process but high aspect ratio is not really a prerequisite in the present application.

In step 3, as shown in **Figure 3c**, the upper surface of the UV-opaque SU-8 resin from step 1 is physically deformed by a mechanical impact produced by the embossing stamp moving downwards at high speed. At the same time, the wafer temperature is tuned down. As a result of this impact, an array of micro-conical wells will be 3D cast on the upper surface of the UV-opaque SU-8. It is important to understand that the stroke speed of stamping will determine the sharpness of the tip of each conical well and the surface profile of the inner wall. In general, if the stroke speed is higher, the sharp tip of each conical well will accordingly become sharper.

In step 5, as shown in **Figure 3d**, the embossing stamp is removed from the wafer while the wafer is being cooled down. Following this cooling step, the conical micro-wells will become highly solidified.

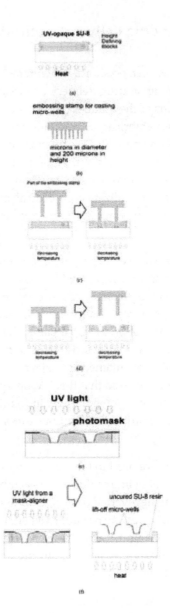

Figure 3. Cross-sectional illustration of the procedure for fabrication of a micro-funnel. (a) A wafer containing a bath of UV-opaque highly dehydrated SU-8 resin. (b) Cross-sectional view that illustrates the embossing stamp with micro-rods. The diameter of each micro-rod is 40 μm. The length of each micro-rod is 200 μm. (c) Casting of micro-wells by surface deformation using the embossing stamp. (d) Removal of the embossing stamp from the wafer. (e) UV exposure. (f) Removal of uncured SU-8 resin by melting at an elevated temperature.

In step 4, as shown in **Figure 3e**, the shell of each micro-funnel is formed and thickened by increasing the dosage of UV exposure. The micro-well patterns on the wafer can be photo-lithographically defined, patterned, and exposed to a UV light using a photomask. Since the SU-8 resin in the wafer contains a UV-opaque impurity, the micro-wells become partially transparent at ultraviolet spectrum. In the presence of the UV-opaque impurity, the interior surface of each micro-well is the only region fully exposed to the UV lights. The SU-8 resin attached to the opposite side of the UV-exposed surface will be partially cured. As a result, the interior surface of each micro-well will be polymerized into a hard and thick layer. The thickness of this polymeric layer can be easily increased by increasing the duration of the UV exposure. In general, this polymeric layer can be as thin as a membrane or as thick as the application demands, depending on the duration of UV exposure.

In step 5, the uncured UV-opaque SU-8 resin attached to each of the micro-well is removed from the micro-well by melting at an elevated temperature. The SU-8 resin attached to the opposite side of the UV-exposed surface will remain uncured and melted into a liquid when the wafer is subjected to a strong heat. As a result of this heat, the UV-exposed surface will be significantly hardened. Traces of the uncured photosensitive resin which remain attached to the micro-well array can be stripped off by developing in 1-methoxy-2-propanol acetate.

In step 6, a hole is formed on the tip of each conical micro-well by dry etching. This step is intended to turn each conical micro-well into a micro-funnel. A hole can be formed on the tip of each conical micro-well by dry-etching the wafer in oxygen plasma for 100 seconds using a Trion RIE/PECVD tool. The oxygen plasma also sharpens the tip of each micro funnel during the dry etching process. In the present study, the process parameters were 90% O2, 10% CF4, an RF power of 100 w, and a chamber pressure of 1.6 Torr.

10. Results and discussions

Figure 4 shows one of the micro-funnels which have been fabricated using the abovementioned method. The micro-funnels were designed to have a sharp tip and wide top, that is a low aspect ratio geometry. This design allows a larger amount of drug to be encapsulated per micro-funnel. The conical geometry used in this study had a volume in excess of 24.4 nl. In addition, the micro-funnels were found to have sufficient mechanical strength for inserting into the

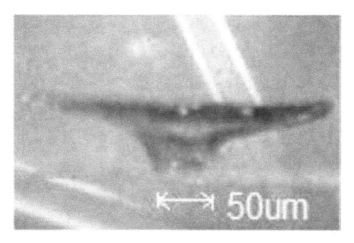

Figure 4. Photo illustrating a fabricated micro-funnel.

human skin. The smallest tip diameter which has been achieved in this study is within 50 μm, which is sharp enough to pierce the human skin.

The proposed fabrication methodology can be further improved and extended to realize other more complicated systems involving 3D micro-structures.

11. Conclusion

For the first time, this article has presented a method that enables concave or convex micro-structures with shells elements or hollow space to be photo-lithographically fabricated. The method comprises four basic stages. The first stage involves preparation of a composite containing a mixture of SU-8 monomer and a UV-opaque impurity. The second stage involves fabrication of an embossing stamp for casting the desired 3D patterns. In the third stage, the desired 3D hollow or suspended micro-structures are cast using the embossing stamp. In the final stage, features requiring with suspended or hollow parts are selectively polymerized by UV exposure. The concepts of this technology have been demonstrated by fabrication of prototypes.

Acknowledgements

This work is supported by Guangdong Natural Science Funds for Distinguished Young Scholar (2015A030306032), National Natural Science Foundation of China (61401191), Shenzhen Science and Technology Innovation Committee funds (JCYJ20150331101823678).

Author details

Louis WY Liu*, Qingfeng Zhang and Yifan Chen

*Address all correspondence to: liaowy@sustc.edu.cn

Department of Electrical Electronics Engineering, South University of Science Technology of China, Shenzhen, China

References

[1] Bertsh A, Lorenz H, Renaud P. Combining microstereolithography and thick resist UV lithography for 3D microfabrication. In: Proceedings of the IEEE International Conf. on

Micro Electro Mechanical Systems (MEMS 98); 25–29 January 1998; Heidelberg, Germany; pp. 18–23.

[2] U.S. Patent 4,575,330. Apparatus for production of three-dimensional objects by stereolithography.

[3] Paar M. Design of injection moulds for thermosets. Dissertation, RWTH Aachen University, 1984.

[4] Spennemann A. A new machine and processing technology for the injection moulding of micro parts. Dissertation, RWTH Aachen University, 2000.

[5] Liu WY. Micro machining process for forming three dimensional micro structures having sloping and profiled side walls. UK Patent Application GB 2404454A, 2005.

[6] Liu WY. Photolithographical micro fabrication of hollow three dimensional structures with profiled side walls. UK Patent Application GB2416599, 2006.

[7] Virdee BS, Liu WY. On-chip cooling of RFICs and MICs. In: 34th European Microwave Conference; 12–14 October 2004; Amsterdam, Netherlands; pp. 173–175.

[8] Bertsch A, Jiguet S, Bernhard P, Renaud P, Pique A, Holmes AS, Dimos DB. Microstereolithography: a review. In: MRS Proceedings. Symposium on Rapid Prototyping Technologies; 3–5 December 2002; Boston, Massachusetts; p. 758, LL1.1. DOI: 10.1557/PROC-758-LL1.1

[9] Rogers JD, Kärkkäinen AHO, Tkaczyk T, Rantala TJ, Descour MR. Realization of refractive microoptics through grayscale lithographic patterning of photosensitive hybrid glass. Optics Express. 2004; 12(7): 1294–1303.

[10] Rammohan A, Prabhat K, Dwivedi RMD, Katepalli H, Madou MJ, Sharma A. One-step maskless grayscale lithography for the fabrication of 3-dimensional structures in SU-8. Sens. Actuators B: Chemical. 2001; 153(1): 125–134. DOI: 10.1016/j.snb.2010.10.021

[11] Chen M, Yao D, Kim B. Optimization of process conditions in gas assisted injection molding. Polym. Plast. Technol. Eng. 2001; 40(4): 479–489. DOI: 10.1081/PPT-100002071

[12] Chen M, Yao D, Kim B. Eliminating flow induced birefringence and minimizing thermally induced residual stresses in injection molding parts. Polym. Plast. Technol. Eng. 2001; 40(4): 491–503. DOI: 10.1081/PPT-100002072

[13] Yao D, Kim B, Choi J, Brown R. Optimizing injection molding toward multiple quality and cost issues. Polym. Plast. Technol. Eng. 1999; 38(5): 955–966. DOI: 10.1080/03602559909351624

[14] Liu WY, Steenson DP, Steer MB. Membrane-supported copper E-plane circuits. In: IEEE MTT-S; 20–24 May 2001; Phoenix, AZ; pp. 539–542.

[15] Liu WY, Steenson DP, Steer MB. Organic micromachining techniques for mass production of millimeter-wave and submillimeter-wave planar circuits. J. Micro/ Nanolith. MEMS MOEMS. 2002; 1(2): 150–153. DOI: 10.1117/1.1463042

[16] Liu WY, Steenson DP, Steer MB. Membrane-supported CPW with mounted active devices. IEEE Microwave and Wireless Components Letters. 2001; 11(4): 167–169. DOI: 10.1109/7260.916332

[17] Liu WY, Mohammadi S, Katehi LPB, Steer MB. Polymer-membrane-supported fin-line frequency multipliers. In: Radio and Wireless Conference (RAWCON 2002); 17–19 August 2002; IEEE; 2002. pp. 281–284.

[18] Liu WY, Mohammad S, Katehi LPB, Khalkhali H, Korabayashi K. Micro-heat-pipe for InP/InGaAs microwave integrated circuits. In: EDMO 2002; 18–19 November 2002; IEEE; 2002. pp. 143–148.

[19] Liu LWY, Virdee BS, Inal J, Steer MB. Microfabrication of conical micro-funnels for drug delivery applications. Micro & Nano Letters. 2015; 10(7): 355–357.

5

Bioengineered Surfaces for Real-Time Label-Free Detection of Cancer Cells

Nicola Massimiliano Martucci, Nunzia Migliaccio,
Immacolata Ruggiero, Ilaria Rea,
Monica Terracciano, Luca De Stefano, Paolo Arcari,
Ivo Rendina and Annalisa Lamberti

Abstract

Biosensing technology is an advancing field that benefits from the properties of biological processes combined to functional materials. Recently, biosensors have emerged as essential tools in biomedical applications, offering advantages over conventional clinical techniques for diagnosis and therapy. Optical biosensors provide fast, selective, direct, and cost-effective analyses allowing label-free and real-time tests. They have also shown exceptional potential for integration in lab-on-a-chip (LOC) devices. The major challenge in the biosensor field is to achieve a fully operative LOC platform that can be used in any place at any time. The choice of an appropriate strategy to immobilize the biological element on the sensor surface becomes the key factor to obtain an applicable analytical tool. In this chapter, after a brief description of the main biofunctionalization procedures on silicon devices, two silicon-based chips that present an (i) IgG antibody or (ii) an Id-peptide as molecular probe, directed against the B-cell receptor of lymphoma cancer cells, will be presented. From a comparison in detecting cells, the Id-peptide device was able to detect lymphoma cells also at low cell concentrations (8.5×10^{-3} cells/μm^2) and in the presence of a large amount of non-specific cells. This recognition strategy could represent a proof-of-concept for an innovative tool for the targeting of patient-specific neoplastic B cells during the minimal residual disease; in addition, it represents an encouraging starting point for the construction of a lab-on-a-chip system for the specific recognition of neoplastic cells in biological fluids enabling the follow-up of the changes of cancer cells number in patients, highly demanded for therapy monitoring applications.

Keywords: biochip, surfaces, biosensing, optical, silicon, idiotype peptide

1. Introduction

Biosensing is a scientific and technological hot topic, given its potential in the field of medical diagnosis [1], healthcare [2], environment [3], defense [4], and food security [5].

Nowadays, the healthcare and pharmaceutical sectors continuously demand more powerful analytical and diagnostic tools for the identification of disease, the development of new medicines, and better diagnostic tests. In these fields, the specific and sensitive detection of targets in short-time analyses plays a key role. While a number of commercial bioassay kits are already on the market, there still remains a major challenge to develop novel biodetection methods to meet the ever-increasing request. High-capability optical biosensing systems are actually emerging as a way to reach this aim [6].

Optical biosensing is a powerful analytical tool used to detect optical changes upon the interaction between an analyte of interest and its ligand, previously immobilized on the sensing device as biological probe; the intensity of obtained signal can be measured, and its value is a function of the analyte concentration in the sample [7, 8]. Generally, optical biosensors use two different detection protocols: direct detection of the analyte (label-free detection) or indirect detection through optically labeled probes (label-based detection). In label-based detection, fluorescent, enzymatic, or radioactive tags, linked on target or on probe molecules, are used; the intensity of the signal indicates the presence of analyte and the interaction with the recognition molecule. This type of detection is extremely sensitive, since it is possible to detect up a single molecule [9, 10]; nevertheless, complex procedures are needed for labeling, and they may interfere with the functionality of the target molecule. In contrast, in label-free detection, target molecules are not labeled or altered and are free to interact in their natural forms: Recent progresses in this field are showing promising results [10]. This strategy is relatively easy and cheap to perform and allows for quantitative and kinetic measurement of molecular interactions. These advantages, with low detection limit, non-destructive approach, and with the ability to recognize a wide variety of analytes or multiple analytes at the same time with fast signal monitoring and analysis [11], make label-free optical detection one of the leading recognition methods in the biosensor field.

Label-free biosensors are very versatile platforms, since they can be developed in different materials, such as silicon or its compounds, glasses, metals, or polymers, and they offer different detection modes and configurations that can be combined [12]. In perspective, optical label-free biosensors are expected to replace fluorescent biosensors in DNA micro-arrays and lab-on-a-chip (LOC) applications [13–15].

The use of silicon-based technology—the same developed for microelectronics—could allow the integration of microfluidic circuits for analyte handling, sensing elements, and control/reading electronics into a single chip. This could pave the way to the mass production of micro total analysis systems and LOCs capable to provide rapid, sensitive, and multiplexed measurements in any place at any time.

The selection of the biological element to immobilize on the sensor surface is dictated by the application and must be chosen to be highly specific for the target molecule and stable enough to be immobilized without losing functionality. Several types of routes can be used to biofunctionalize the sensor surface [16], and the choice of an appropriate immobilization procedure has become a key factor in the biosensor area. An ideal immobilization procedure should guarantee an efficient coverage by the molecular probes of the sensing layer without interfering with their properties (structure, biological activity, affinity, specificity). In addition, the possibility to preserve their stability in normal storage conditions and also for regeneration could be useful in the case of integration in portable LOC. Despite enormous research efforts have been made to find novel strategies according to the application, a universally valid procedure has not yet been developed and the realization of cheap hand-held platforms is almost limited. For this reason, the choice of the most effective strategy of immobilization represents the critical step that turns a sensing device into an applicable analytical tool with the required quality standards. Currently, a lot of biomolecules can be used as bioreceptors (antibodies, nucleic acid, peptides, enzymes, cell receptors, and many others). Among these, artificial peptides provide an opportunity to develop the desired molecular biosensor due to their desirable properties such as diversified structure, high affinity to ligands, matured synthesis protocol, and modified approach [17].

2. Biosensing application in cancer

Cancer is one of the main causes of death worldwide. Early diagnosis is the key to enhance the success of medical treatment. In the last few years, in parallel with a growing interest in detecting cancer cells, a wide variety of techniques were developed for detection at the cellular level [18–21]; nevertheless, most of these modalities are expensive and time-consuming, and they are often associated to risks deriving from radioactive tracers.

At this point, despite some considerable achievements, the realization of simple, rapid, non-destructive, and low cost methods for early detection of cancer and minimal residual disease, important for diagnosis and reduction in mortality for certain cancers, still remains an unfulfilled goal [22, 23]. To meet these specific requirements, biosensors have attracted increasing attention since biosensing technology, taking advantage of the properties of biological systems combined to functional advanced materials, is providing rapid, reproducible, and highly sensitive cell detection.

Among the various types of cancer, lymphoma is the most common blood cancer, which incidence is recently increasing. This malignancy is a clonal expansion of neoplastic cells that may result in fatal outcomes [24]. Despite the great progress that has been made over the last several decades in the treatment of lymphoma, the prognosis for patients with particular sub-types of disease remains quite poor. Lymphomas are a heterogeneous group of tumors arising in the reticuloendothelial and lymphatic systems. The major types are Hodgkin lymphoma and non-Hodgkin lymphoma. Great progresses in the use of monoclonal antibodies, chemotherapy, radioimmunotherapy, and peripheral blood stem cell transplants have achieved significant responses in the treatment of these diseases and also markedly improved the outcome of the cure among elected sub-populations of patients, since not all the patients or subtypes of lymphoma are responsive to these "conventional" approaches. For example, tumorigenic B-cell lymphomas are sensitive to anticancer treatments, including convention-al chemotherapy, radiation therapy, and corticosteroids [25]. Nevertheless, the disease is associated with incomplete response to clinical treatments that result in a minimal residual disease in which a few neoplastic cells undetected *in vivo* replenish the cancer cell reservoir. This grim scenario calls for novel strategies to detect tumorigenic B cells. Random peptide libraries (RPLs) allow the selection of therapeutic peptides for tumor cell-surface receptors. Idiotypic determinants of the Ig-BCR, expressed by lymphoma cells, function as a specific tumor antigen that may be exploited for cell-specific detection or targeted therapy. Here, we present the complex different strategies that we adopted in order to construct a biosensor for the detection of tumorigenic lymphoma B cells and discuss all the difficulties that we encoun-tered and the approaches we adopted for their overcoming.

3. The choice of appropriate support for biosensing

Our vision was to create a highly sensitive, label-free optical biosensing system for the targeting of patient-specific neoplastic B cells during the minimal residual disease. The material used to develop this unique tool for biosensing include mainly silicon, since it possesses great potential because of its many unique properties, including biocompatibility, which is an important precondition for biological and biomedical applications, abundance, tailorable surface chemistry, and unique electronic, optical, photonic, and mechanical properties, among others. In addition, high surface-to-volume ratio of silicon derivatives offers exciting opportunities to design high-performance silicon-based functional devices for biomedical applications. Moreover, silicon is very abundant on earth allowing inexpensive resources for large-scale and low-cost preparation for practical applications. Taking advantage of these attractive features, the interest in the use of silicon is widely grown leading to its applications not only in biology but also in a lot of other fields [26].

In our research activity, flat silicon devices of fixed thickness (400 μm) were obtained from highly doped p$^+$ type, 0.003 Ω cm resistivity, (100)-oriented silicon wafers, cut into 10 × 10 mm square pieces. The wafers were cleaned by a standard RCA process [27] and thermally oxidized at 1050°C for 5 h. An electrochemical etching process of planar silicon was used to pattern porous silicon with a high specific surface (up to 500 m^2/cm^3). An advantage in the use of porous silicon is that its morphology can be tuned by modification of process parameters [28] so that the resulting structures can be adapted to obtain the best performance for chemical and biological processes that happen on their surface. Moreover, the porosity of the material coupled to the low-cost production makes porous silicon an ideal bulky model system to follow each functionalization step: The concentration of reagents and molecules in the pores allow to quantify few nanometers thick films of passivating agents, exploiting signal enhancements, that cannot easily measured on flat supports.

4. Chemical functionalization procedures

The selection of an appropriate procedure for the immobilization of a biological element on the sensor surface that interacts with the desired target for the specific recognition of an analyte has become a critical step in the biosensor area, and enormous efforts are continuously invested in order to optimize novel strategies according to the application. The biofunctionalization of chemical modified surfaces can be achieved in several manners that can be grouped in just two approaches: (i) direct adsorption and (ii) physical adsorption [29–33]: in both cases, each immobilization route presents advantages and drawbacks.

In the direct adsorption method, there is no bond formation between probe and device, reagents are not required so either structure or functionality of biomolecules is not affected. Nevertheless, the efficiency of this strategy is very low.

The bioreceptor physisorption is a quick and widely used approach to immobilize biomolecules on chip surfaces based on electrostatic, hydrophobic, and covalent interactions. Despite the efficiency and the simplicity, electrostatic, and hydrophobic approaches are direct fast methods, since no linker molecules are needed, but are limited to situations that require no directional orientation of the bioprobes. Moreover, both techniques request long incubation times and the risk of folding and desorption due to changes of parameters, such as pH, ionic strength, or temperature, cannot be excluded. The covalent attachment of probe is more efficient in terms of stability and binding strength. Generally, the binding occurs between a functional chemical group of the biomolecules, whose blocking does not affect the functionality, and one on the modified surface. For proteins covalent coupling, amino, carboxylic, or thiol groups are preferred, whereas in the case of nucleic acids, it is possible to take advantage of the versatility of their synthesis to insert reactive groups at the end of the sequence. More difficult is the immobilization of immunoglobulins in a correct orientation, which can be achieved by controlled linkage of carbohydrates groups in the constant region or using affinity proteins (such as A or G Protein) [31]. In all physical adsorption types, a chemical modification of the platform surface is required to the extent that the material properties are tuned to accomplish the best analytical characteristics.

The drawing up of an efficient and correct immobilization procedure is a crucial point to avoid a wide variety of factors that may negatively affect the biosensor functionality. The orientation of probe, the density of coating on the detection surface, pH, target concentration, operating conditions, and chemical environment provided by transducer must be closely explored. An efficient biofunctionalization process should take in account few important observances: The preservation of the molecular probe structure to guarantee subsequent binding of analyte; limited chemical steps and minimal consumption of reagents and samples to make the whole procedure lean and easily reproducible; low optical adsorption at the working wavelengths; homogeneously thin layer formation compatible with evanescent field sensing; uniform surface coating; saturation of eventually free binding sites to reduce the possibility of false-positive signals; biocompatible conditions; integrability with large-scale fabrication. The exploration of these traits offers the possibility to improve biosensors features increasing the power of detection.

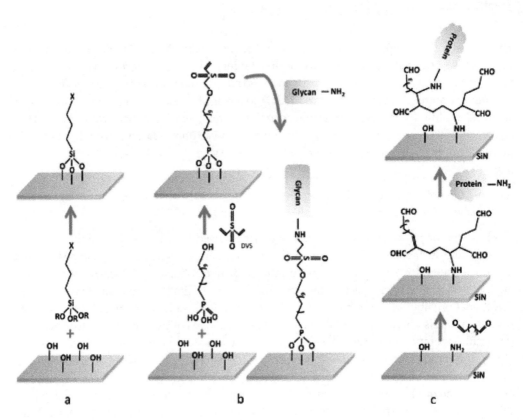

Figure 1. Types of silicon surface chemical modifications for biosensors: (a) organosilane-based, (b) phosphonate-based, and (c) glutaraldehyde-based strategies.

Figure 1 represents the main functionalization approaches employed to construct integrated optics (IO) biosensors. Before the biofunctionalization step, a previous chemical activation of the sensor surface is always needed. To this aim, our group employed the self-assembly of organofunctional alkoxysilanes (**Figure 1a**), an easy and versatile system for organic conjugating [34]. However, silicon-based surfaces require a prior activation step to oxidize the surface and to expose the silanol groups for cross-linking with the silane. The formation of a thin silane self-assembled film allows applying a great number of chemical reactions. Immediately before silanization, surfaces are cleaned with oxidant media to remove organic pollutants and to increase the hydroxyl moieties on the surface [35]. The used oxidant is piranha solution [36–39], consisting of a concentrated sulfuric acid mixed with hydrogen peroxide at 3:1 ratio. This treatment is performed by heating for 30 min only. Hundreds of different organosilanes with different structures and functionalities are nowadays commercially available, although the most commonly employed are those with short alkyl chain that present an amino, thiol, epoxy, or carboxylic group at the terminus. Among this vast variety of compounds, 3-aminopropyltriethoxysilane (APTES) was chosen for its reactivity to aldehyde, carboxylic acid, and epoxy functionalities.

The reaction between the oxidized surface and the organosilane is based on the condensation between the Si–O–Si of the silane and the OH present on the device; generally, besides the hydroxyl groups already present on the native silicon oxide layer, a thermal oxidation is a common procedure to form a new efficient oxide film [40–42] in order to assure a plenty of silanol groups for an efficient coverage of the organic layer.

Furthermore, after silanization, APTES layer was cured at high temperature [43]. The aminosilane is more reactive, and it can be applied on a surface using pure organic solvent. The advantage of the curing is that a more controlled deposition of the compound can be obtained to create a thinner film of the aminopropyl groups on the chip. The self-assembled monolayers generally present a thickness in the range of 1–3 nm and create a nanometer-scale organic thin coat [44, 45].

5. Choice of biomolecular probe

At this point, a wide variety of biomolecules (antibodies, nucleic acid sequences, peptides, enzymes, cell receptors) can be used as bioreceptors (**Figure 2**).

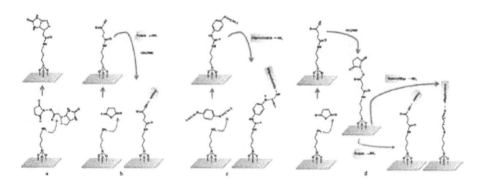

Figure 2. Types of bioconjugation methods on aminated surfaces: (a) N-hydroxysuccinimide–based, (b) succinic anhy-dride–based, (c) p-phenylenediisocyanate–based, and (d) glutaraldehyde-based strategies.

The choice of bioreceptor depends on the intended application of biosensor and it must meet two important requirements: high specificity for the target molecule and high stability to retain its biological activity when immobilized on the support.

A first biofunctionalization approach, based on the covalent bind of a biomolecule on the activated silicon sensor surface, included the use of an IgG antibody as molecular probe directed against B-cell receptor. The chip was treated with the homobifunctional cross-linker glutaraldehyde (GA): This molecule, besides to be employed to form an aldehyde-terminated surface, which allows the reaction of amine groups, by the formation of imines (Schiff bases), acts as spacer in order to keep away from the surface the immobilized bioprobe that can react freely with target molecules [46, 47]. By this strategy, the antibody has been immobilized on the surface via protein A in an oriented fashion [48]. The whole process is checked monitoring surface changes by ellipsometric measurements and FTIR spectromicroscopy. As reported in Table 1, using a random sampling of four different wells, it was observed for all of them after each functionalization step the surface layer thickness.

	Thickness (nm)			
Film	Sample 1	Sample 2	Sample 3	Sample 4
Oxide	75.8 ± 0.4	72.4 ± 0.2	75.0 ± 0.3	73.4 ± 0.2
Aptes + GA	3.0 ± 0.4	3.1 ± 0.3	3.1 ± 0.3	2.1 ± 0.2
Protein A	0.68 ± 0.09	0.67 ± 0.08	0.75 ± 0.1	0.85 ± 0.1
χ^2	0.54	0.43	0.47	0.45

Table 1. Surface layer thickness on four random samples after each step of functionalization measured by ellipsometry technique.

The analyses of the FT-IR spectra led to the identification of several characteristic vibration bands that were coherent with the various functionalization steps. Table 2 reports a list of the major bands identified together to peak assignment.

Predicted peak	Frequency cm⁻¹					
	Sample 1	Sample 2	Sample 3	Sample 4	Sample 5	Sample 6
Si–O str	–	1127	–	–	–	–
Si–O–C *as* str	–	–	1250	1250	1258	1258
–(CH₂)–str	–	–	1295–1305	–	–	–
–O–CH₂–str	–	–	1445–1475	–	–	–
Amide II	–	–	–	–	–	1531
C=O str	–	–	–	–	1635	1642
Saturated primary ammine (–NH₂ def)	–	–	1650	–	–	–
Amide I C=O str	–	–	–	–	1650–1680	1638–1687
C=O str	–	–	–	–	1685–1705	1774
N–H str	–	–	–	–	–	3121
Primary ammine –NH₂ str	–	–	3250–3677	–	–	–

Table 2. Major bands identified by FTIR spectromicroscopy and corresponding assigned peaks.

As experimental model, it was chosen a murine lymphoma cell line (A20) [49] that expresses high levels of membrane IgG. The most interesting point of this first approach is that the microfabricated biochip appears to be suitable to reveal specific bindings such as that between cell-surface proteins (receptor) and corresponding specific antibody. In addition, the number of cells detected by the devices was 2.0×10^{-3} cells/μm^2.

Anyway, since this detection limit does not seem satisfactory and the idea that the contact probability between cells and antibodies on capture specific surface could be improved, we took advantage of a new functionalization strategy exploiting an Id-peptide as biomolecular probe. The choice of an Id-peptide was dictated by two main reasons: (i) Artificial peptides provide an opportunity to develop the desired molecular biosensor due to their desirable properties such as diversified structure and high affinity [50]. In addition, peptides with specific sequences can provide high affinity to particular ligands and be obtained by screening and optimization of artificial peptide libraries; (ii) the used idiotype peptide is a small peptide ligand able to be recognized with high affinity and specificity from the B-cell receptors present on the lymphoma B cells [51–53]. The use of a small ligand as biorecognition element endowed with great specificity could highly enhance affinity and selectivity of the detection layer. In addition, it simplifies the functionalization procedure with respect to that employed for antibodies in which controlling protein orientation is still very challenging [54]. The peptide was immobilized on the silicon surface following the functionalization strategy schematized in **Figure 3**.

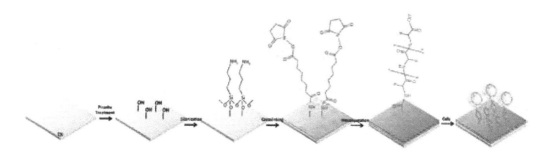

Figure 3. Functionalization approach utilized on silicon surface to conjugate an Id-peptide to detect lymphoma cells. After each passivation step, the new synthesized layer is reported in the figure with the same color of the molecule used in the chemical reaction (APTES is blue, BS³ is red, Id-peptide is green).

This chemical procedure was developed on both crystalline flat and porous silicon samples; the nanostructured porous was chosen because its peculiar morphology allows the immobilization of a greater number of molecules with respect to a planar substrate and a number of functionalization investigation methods could be more easily exploited [55]. The aminosilanized surface has been activated by the homobifunctional cross-linkers bis[sulfosuccinimidyl] suberate (BS³), which, acting as spacer, provide succinimidyl-activated carboxyl group that could react with amine-ended peptide to form an amide bond. Changes in chemical composition of PSi surface were monitored by FTIR spectroscopy after each functionalization step until BS³ (**Figure 4**).

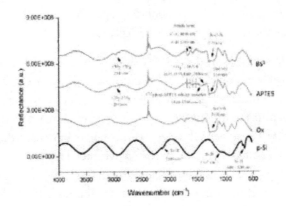

Figure 4. FTIR spectra of silicon surface after each chemical modification step.

The analysis of the FTIR spectra in the range from 2500 to 500 cm^{-1} highlighted characteristic peaks of each molecule used in the different passivation steps, demonstrating the effectiveness of the functionalization procedure. Indeed, the characteristic peaks of Si–Hx bonds corresponding to the PSi after electrochemical etching (2100 and 680–630 cm^{-1}) are no longer visible when the devices were thermally oxidized, whereas the appearance of the Si–O–Si characteristic band at 1100 cm^{-1} was detected.

The formation of the silane film was confirmed by the presence of peaks in the span 1440–1390 cm^{-1}, relative to CH_3 from APTES ethoxy moieties, and at 1655 cm^{-1} relative to an imine group from oxidation of an amine bicarbonate salt [56]. Moreover, the appearance of the peaks at 1640 and 1550 cm^{-1} that correspond to CO– and NH– groups of an amide bond, confirms the deposition of the BS^3.

The functionalization of porous silicon surface was also confirmed by spectroscopic reflectometry.

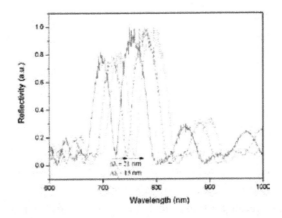

Figure 5. Reflectivity spectra on porous silicon surface before (solid line) and after APTES silanization (dashed line), and after BS^3 functionalization (short dashed line).

Reflectivity spectra of porous silicon devices during functionalization steps are reported in **Figure 5**. The deposition on pores walls of a thin layer, constituted by the different organic chemical compounds, produces red shifts of spectra due to the increase of the average refractive index of porous silicon surfaces [57]. After silanization and cross-linker modification, a red shift of 21 and 15 nm was recorded, respectively. The same chemical modifications were performed also on flat silicon devices; in the latter, the whole functionalization procedure was followed by spectroscopy ellipsometry, in order to quantify layer thickness variations after thermal oxidation (SiO_2), silanization (APTES), and cross-linker functionalization (BS^3). As showed in Table 3, the thickness of oxidized silicon devices was 74 ± 1 nm; this value increased of 3 and about 2 nm after treatment with aminosilane and BS^3, respectively.

	Thickness (nm)			
Film	**Sample 1**	**Sample 2**	**Sample 3**	**Sample 4**
Oxide	74.9 ± 0.3	74.9 ± 0.2	73.5 ± 0.2	74.7 ± 0.3
Aptes	3.5 ± 0.2	2.9 ± 0.3	3.1 ± 0.2	2.7 ± 0.4
BS^3	1.72 ± 0.08	1.84 ± 0.03	1.82 ± 0.03	1.78 ± 0.02

The values reported are the average of five determinations on each sample.

Table 3. Surface layer thickness on four random samples after each step of functionalization measured by ellipsometry technique.

6. Biosensing

Once the chemical modified silicon chips have been obtained, a procedure to immobilize a small peptide for label-free detection of cancer cells was settled. The used experimental system takes advantage of the properties of an idiotype peptide isolated from peptide libraries able to bind the variable region of the B-cell receptor on A20 lymphoma cells [51]. The selected peptide, named A20-36 (pA20-36), whose sequence is EYVNCDNLVGNCVI, was linked on silicon-modified surfaces and used as molecular probe. A random peptide (RND), SSAYGSCKGPCSSGVHSI, was used as negative control. To determine the optimal peptide concentration to obtain a uniform coverage of planar and porous surfaces, a titration was carried out. Based on the obtained results [58], 150 µM concentration was used for both peptides.

The detection of lymphoma cancer cells fulfilled on both planar and porous peptide-modified silicon surfaces is showed in **Figure 6**. The panels *a* and *b* report microscope light images of the planar device surfaces after incubation with a low (100 cells) or high number (50,000 cells) of A20 cells. The choice of the high number of cells was made in order to have saturation binding conditions. The same number of cells (50,000) was incubated on A20-36-peptide-modified porous silicon surface, but a lower number of detected cells were observed on light microscope (**Figure 6**, panel *c*). The chip was not able to bind lymphoma cells when functionalized with RND peptide (**Figure 6**, panel *d*), whereas no myeloma cells (5T33MM), a surface IgG-positive B-cell line unable to bind to pA20-36 peptide [51], were detected when incubated on the device functionalized with pA20-36 (**Figure 6**, panel *e*).

The number of A20 cells detected on functionalized planar surface device was about 8500 and, taking in account an average area of 80 μm^2 for a single cell, filled ~680,000 μm^2, a value concordant with the available functionalized area (~1.0×10^6 μm^2); when the detection was performed on porous silicon device, the number of cells that effectively bind the chip was lower (400), filling an area of about 32,000 μm^2. The exiguous number of A20 cells on the porous silicon surface was probably caused by the peculiar morphology of the support; being highly porous, with pore diameter of about 50 nm, and pore upper edges lower than 1 nm in thickness, its inner surface is many order of magnitude greater than the top active one. Hence, just a very low number of peptides are really available on the pore upper edges to bind the cells (that cannot enter into the pores). Therefore, the consequence of this condition is the decrease in the number of cells detected on porous silicon biochip resulting lower respect to that on the planar surface.

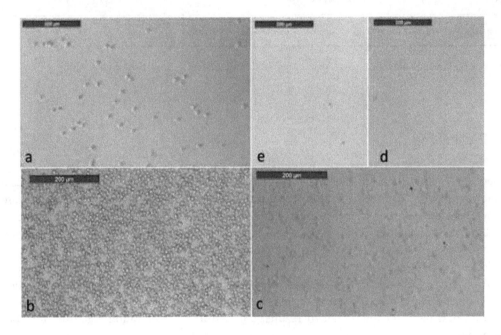

Figure 6. Optical images of A20 cell detection on both planar and porous silicon devices. Planar silicon pA20-36 modified sensor after incubation with 1×10^4 A20 cells/mL (*a*) and 5×10^6 A20 cells/mL (*b*). Porous silicon pA20-36 modified surface after incubation with 5×10^6 A20 cells/mL (*c*). Planar silicon RND-modified-sensor after incubation with 5×10^6 A20 cells/mL (*d*). Planar silicon pA20-36 modified sensor after incubation with 5×10^6 5T33MM cells/mL (*e*).

The surface of each silicon chip presents a functionalized available area of about 1.0×10^6 μm^2 so the maximum number of cells that can be bound on the device was 10,000 (covering an area of about 800,000 μm^2). Since the number of cells detected on planar and porous silicon surfaces was by count 8500 and 400 (evaluated by optical microscopy), the efficiency of detection is 85 and 4%, respectively. Moreover, comparing the efficiency of detection of the flat silicon device based on Id-peptide-BCR recognition with an analogous silicon-based bioanalytical system in which an anti-IgG-BCR was used as molecular probe [48], it is clear that the first biochip resulted more efficient in detecting A20 cells (8.5×10^{-3} vs. 2.0×10^{-3} cells/μm^2, respectively).

This difference is likely due to the better accessibility of the A20-36 Id-peptide on the BCR with respect to the anti-IgG. In fact, the binding of the idiotype peptide should occur with the more exposed variable region of the receptor in contrast with the interaction between IgG and BCR in which the variable regions of the immunoglobulin bind the less exposed constant region of the receptor. Furthermore, also the difference in affinity constants between the two ligand-receptor systems coupled to diverse functionalization approaches might have had a decisive role in the detection efficiency.

Cell detection was also investigated by atomic force microscopy analysis (AFM) (**Figure 7**).

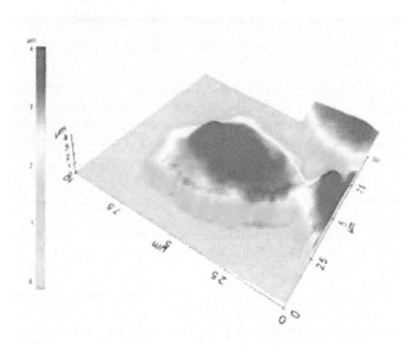

Figure 7. Representative AFM image in the trace direction of live A20 cells detected on silicon surface.

Both light microscopy and AFM analysis showed a good biocompatibility of substrate since viability and cell morphology were not affected.

At this point, since cancerous cells are coexisted with other cell types in the body and it is very important to selectively differentiate cancer cells from other ones, in order to assess the real performance of the biochip, lymphoma cells detection was carried out on devices incubated with mixed samples of A20 and 5T33MM cells (3.5×10^5/mL). The detection of lymphoma cells in system mixed is reported in **Figure 8**.

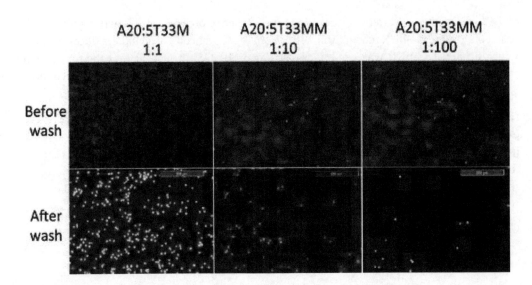

Figure 8. Detection of A20 cells (green) in system mixed with 5T33MM cells (red) by fluorescence macroscopy after incubation on planar silicon pA20-36 modified sensor.

The mixed system has been prepared in three different ratios (A20:5T33MM = 1:1, 1:10, 1:100) of the two labeled live cell lines [59]. The efficiency of detection also in a complex system demonstrated the high selectivity of the device, confirming that the use of an Id-peptide immobilized on a silicon-based chip could be a good proof-of-concept for future researches.

7. Conclusions

In this chapter, we focused on the functionalization and activation of crystalline and porous silicon surfaces to develop devices allowing the identification of specific ligand-receptor interactions.

As an example, we report new results about the realization of devices suitable to highlight the specific interaction between cell surface receptors and corresponding specific ligands. One of these devices was applied to detect the binding of extremely aggressive murine A20 lymphoma cells to a specific IgG antibody as molecular probe directed against B-cell receptor. The result was encouraging and prompted us to develop an improved device, more sensitive, for the specific recognition of different types of tumor cells. Another approach was based on the specificity of an idiotype peptide endowed with high-affinity toward A20 lymphoma cells. Particularly, the use of an Id-peptide as probe allowed to obtain a uniform sensor surface coating, thus enhancing capture ability also at low cell concentrations. Moreover, the biosensor was biocompatible and showed high repeatability as well as selectivity in label-free cell detection.

The improved device opens the way to the development of unique diagnostic tools in point-of-care testing for recognition and isolation of patient-specific neoplastic B cells during the minimal residual disease. Any idiotype peptide is ideally endowed with a unique, clone-specific antigenic reactivity. Of course, this approach requires the selection of Id-peptides for each patient through laborious and costly procedures. This might be overcome focusing on a specific B-cell tumors, where a consistent number of patients share the same antigenic reactivity against a restrict pool of Id-peptides. Nevertheless, this strategy can be utilized for the characterization of other specific peptide–receptor interactions through the screening of a recombinant phage library.

Acknowledgements

This work was supported by funds from Programmi di Ricerca Scientifica di Rilevante Interesse Nazionale (2012CK5RPF_004) and POR Campania FSE 2007–2013, Project CRÈME. The authors thank Dr. P. Dardano from IMM-CNR for AFM analysis and Prof. G. Scala and C. Palmieri from University "Magna Graecia" for providing Id-peptides.

Author details

Nicola Massimiliano Martucci[1*], Nunzia Migliaccio[1], Immacolata Ruggiero[1], Ilaria Rea[2], Monica Terracciano[2], Luca De Stefano[2], Paolo Arcari[1], Ivo Rendina[2] and Annalisa Lamberti[1]

*Address all correspondence to: nicolamassimiliano.martucci@unina.it

1 Department of Molecular Medicine and Medical Biotechnology, University of Naples, Federico II, Naples, Italy

2 Institute for Microelectronics and Microsystems of Naples, National Research Council, Naples, Italy

References

[1] McPartlin DA, O'Kennedy RJ. Point-of-care diagnostics, a major opportunity for change in traditional diagnostic approaches: potential and limitations. Expert Rev Mol Diagn. 2014; 14:979–998. doi:10.1586/14737159.2014.960516

[2] Kumar S, Ahlawat W, Kumar R, Dilbaghi N. Graphene, carbon nanotubes, zinc oxide and gold as elite nanomaterials for fabrication of biosensors for healthcare. Biosens Bioelectron. 2015; 70:498–503. doi:10.1016/j.bios.2015.03.062

[3] Bereza-Malcolm LT, Mann G, Franks AE. Environmental sensing of heavy metals
     through whole cell microbial biosensors: a synthetic biology approach. ACS Synth Biol.
     2015; 4:535–546. doi:10.1021/sb500286r

[4] Upadhyayula VK. Functionalized gold nanoparticle supported sensory mechanisms
     applied in detection of chemical and biological threat agents: a review. Anal Chim
     Acta. 2012; 715:1–18. doi:10.1016/j.aca.2011.12.008

[5] Verma N, Bhardwaj A. Biosensor technology for pesticides: a review. Appl Biochem
     Biotechnol. 2015; 175:3093–3119. doi:10.1007/s12010-015-1489-2

[6] Fan X, White IM, Shopova SI, Zhu H, Suter JD, Sun Y. Sensitive optical biosensors for
     unlabeled targets: a review. Anal Chim Acta. 2008; 620:8–26. doi:10.1016/j.aca.
     2008.05.022

[7] Holford TRJ, Davis F, Higson SPJ. Recent trends in antibody based sensors. Biosens
     Bioelectron. 2012; 34:12–24. doi:10.1016/j.bios.2011.10.023

[8] Turner APF. Biosensors: sense and sensibility. Chem Soc Rev. 2013; 42:3184–3196. doi:
     10.1039/c3cs35528d

[9] Karunakaran C, Bhargava K, Benjamin R, editors. Biosensors and bioelectronics. 1st
     Ed, Elsevier, Netherlands, 2015; p. 1.5.2.

[10] Armani AM, Kulkarni RP, Fraser SE, Flagan RC, Vahala KJ. Label free, single-mole-
     cule detection with optical microcavities. Science. 2007; 317:783–786. doi:10.1126/
     science.1145002

[11] Brogan KL, Walt DR. Optical fiber-based sensors: application to chemical biology. Curr
     Opin Chem Biol. 2005; 9:494–500. doi:10.1016/j.cbpa.2005.08.009

[12] George S, Block ID, Jones SI, Mathias PC, Chaudhery V, Vuttipittayamongkol P, Wu
     HY, Vodkin LO, Cunningham BT. Label-free prehybridization DNA microarray
     imaging using photonic crystals for quantitative spot quality analysis. Anal Chem.
     2010; 82:8551–8557. doi:10.1021/ac101551c

[13] Estevez MC, Alvarez M, Lechuga LM. Integrated optical devices for lab-on-a-chip
     biosensing applications. Laser Photonics Rev. 2011; 6:463–487. doi:10.1002/lpor.
     201100025

[14] Washburn AL, Bailey RC. Photonics-on-a-chip: recent advances in integrated wave-
     guides as enabling detection elements for real-world, lab-on-a-chip biosensing
     applications. Analyst. 2011; 136:227–236. doi:10.1039/c0an00449a

[15] Capitan-Vallvey LF, Palma AJ. Recent developments in handheld and portable
     optosensing—a review. Anal Chim Acta. 2011; 696:27–46. doi:10.1016/j.aca.2011.04.005

[16] Banuls MJ, Puchades R, Maquieira A. Chemical surface modifications for the develop-
     ment of silicon-based label-free integrated optical (IO) biosensors: a review. Anal Chim
     Acta. 2013; 777:1–16. doi:10.1016/j.aca.2013.01.025

[17] Pazos E, Vazquez O, Mascarenas JL, Vazquez ME. Peptide-based fluorescent biosensors. Chem Soc Rev. 2009; 38:3348–3359. doi:10.1039/b908546g

[18] Brindle K. New approaches for imaging tumour responses to treatment. Nat Rev Cancer. 2008; 8:94–107. doi:10.1038/nrc2289

[19] Fass L. Imaging and cancer: a review. Mol Oncol. 2008; 2:115–152. doi:10.1016/j.molonc.2008.04.001

[20] Mosmann T. Rapid colorimetric assay for cellular growth and survival: application to proliferation and cytotoxicity assays. J Immunol Methods. 1983; 65:55–63. doi: 10.1016/0022-1759(83)90303-4

[21] Zhang J, Campbell RE, Ting AY, Tsien RY. Creating new fluorescent probes for cell biology. Nat Rev Mol Cell Biol. 2012; 3:906–918. doi:10.1038/nrm976

[22] de la Rica R, Thompson S, Baldi A, Fernandez-Sanchez C, Drain CM, Matsui H. Label-free cancer cell detection with impedimetric transducers. Anal Chem. 2009; 81:10167–10171. doi:10.1021/ac9021049

[23] Perfézou M, Turner A, Merkoçi A. Cancer detection using nanoparticle-based sensors. Chem Soc Rev. 2012; 41:2606–2622. doi:10.1039/c1cs15134g

[24] Nogai H, Dörken B, Lenz G. Pathogenesis of non-Hodgkin's lymphoma. J Clin Oncol. 2011; 29:1803–1811. doi:10.1200/JCO.2010.33.3252

[25] Migkou M, Dimopoulos MA, Gavriatopoulou M, Terpos E. Applications of monoclonal antibodies for the treatment of hematological malignancies. Expert Opin Biol Ther. 2009; 9:207–220. doi:10.1517/14712590802650581

[26] Peng F, Su Y, Zhong Y, Fan C, Lee ST, He Y. Silicon nanomaterials platform for bioimaging, biosensing, and cancer therapy. Acc Chem Res. 2014; 47:612–623. doi: 10.1021/ar400221g

[27] Kern W, editor. Handbook of semiconductor wafer cleaning technology: science, technology, and applications. William Andrew Publishing/Noyes; 1993.

[28] Dubei RS. Electrochemical fabrication of porous silicon structures for solar cells. Nanosci Nanoeng. 2013; 1:36–40. doi:10.13189/nn.2013.010105

[29] Tang H, Chen J, Yao S, Nie L, Deng G, Kuang Y. Amperometric glucose biosensor based on adsorption of glucose oxidase at platinum nanoparticle-modified carbon nanotube electrode. Anal Biochem. 2004; 331:89–97. doi:10.1016/j.ab.2004.05.005

[30] Wong LS, Khan F, Micklefield J. Selective covalent protein immobilization: strategies and applications. Chem Rev. 2009; 109:4025–4053. doi:10.1021/cr8004668

[31] Prieto-Simín B, Campàs M, Marty JL. Biomolecule immobilization in biosensor development: tailored strategies based on affinity interactions. Protein Pept Lett. 2008; 15:757–763. doi:10.2174/092986608785203791

[32] De Vos K, Girones J, Popelka S, Schacht E, Baets R, Bienstman P. SOI optical microring resonator with poly(ethylene glycol) polymer brush for label-free biosensor applications. Biosens Bioelectron. 2009; 24:2528–2533. doi:10.1016/j.bios.2009.01.009

[33] Kim DC, Kang DJ. Molecular recognition and specific interactions for biosensing applications. Sensors (Basel). 2008; 8:6605–6641. doi:10.3390/s8106605

[34] Huie JC. Guided molecular self-assembly: a review of recent efforts. Smart Mater Struct. 2003; 12:264–271. doi:10.1088/0964-1726/12/2/315

[35] Aswal DK, Lenfant S, Guerin D, Yakhami JV, Villaume D. Self assembled monolayers on silicon for molecular electronics. Anal Chim Acta. 2006; 568:84–108. doi:10.1016/j.aca.2005.10.027

[36] Pal S, Guilermain E, Sriram R, Miller BL, Fauchet PM. Silicon photonic crystal nanocavity-coupled waveguides for error-corrected optical biosensing. Biosens Bioelectron. 2011; 26:4024–4031. doi:10.1016/j.bios.2011.03.024

[37] Byeon JY, Limpoco FT, Bailey RC. Efficient bioconjugation of protein capture agents to biosensor surfaces using aniline-catalyzed hydrazone ligation. Langmuir. 2010; 26:15430–15435. doi:10.1021/la1021824

[38] Scheler O, Kindt JT, Qavi AJ, Kaplinski L, Glynn B, Barry T, Kurg A, Bailey RC. Label-free, multiplexed detection of bacterial tmRNA using silicon photonic microring resonators. Biosens Bioelectron. 2012; 36:56–61. doi:10.1016/j.bios.2012.03.037

[39] García-Rupérez J, Toccafondo V, Banuls MJ, García Castelló J, Griol A, Peransí-Llopis S, Maquieira A. Label-free antibody detection using band edge fringes in SOI planar photonic crystal waveguides in the slow-light regime. Opt Express. 2010; 18:24276–24286. doi:10.1364/OE.18.024276

[40] Rong G, Najmaie A, Sipe JE, Weiss SM. Nanoscale porous silicon waveguide for label-free DNA sensing. Biosens Bioelectron. 2008; 23:1572–1576. doi:10.1016/j.bios.2008.01.017

[41] De Tommasi E, De Stefano L, Rea I, Di Sarno V, Rotiroti L, Arcari P, Lamberti A, Sanges C, Rendina I. Porous silicon based resonant mirrors for biochemical sensing. Sensors (Basel). 2008; 8:6549–6556. doi:10.3390/s8106549

[42] Massad-Ivanir N, Shtenberg G, Tzur A, Krepker MA. Segal E. Engineering nanostructured porous SiO_2 surfaces for bacteria detection via "direct cell capture". Anal Chem. 2011; 83:3282–3289. doi:10.1021/ac200407w

[43] Rea I, Martucci NM, De Stefano L, Ruggiero I, Terracciano M, Dardano P, Migliaccio N, Arcari P, Taté R, Rendina I, Lamberti A. Diatomite biosilica nanocarriers for siRNA transport inside cancer cells. Biochim Biophys Acta. 2014; 1840:3393–3403. doi:10.1016/j.bbagen.2014.09.009

[44] Love JC, Estroff LA, Kriebel JK, Nuzzo RG, Whitesides GM. Self-assembled monolayers of thiolates on metals as a form of nanotechnology. Chem Rev. 2005; 105:1003–1169. doi:10.1021/cr0300789

[45] Howarter JA, Youngbood JP. Optimization of silica silanization by 3-aminopropyltriethoxysilane. Langmuir. 2006; 22:11142–11147. doi:10.1021/la061240g

[46] Peeters S, Stakenborg T, Reekmans G, Laureyn W, Lagae L, Van Aerschot A, Van Ranst M. Impact of spacers on the hybridization efficiency of mixed self-assembled DNA/ alkanethiol films. Biosens Bioelectron. 2008; 24:72–77. doi:10.1016/j.bios.2008.03.012

[47] Cooper J, Cass T, editors. Biosensors. A pratical approach. 2nd Ed, Oxford University Press, United Kingdom, 2004.

[48] Lamberti A, Sanges C, Migliaccio N, De Stefano L, Rea I, Orabona E, Scala G, Rendina I, Arcari P. Silicon-based technology for ligand-receptor molecular identification. J At Mol Opt Phys. 2012; 948390. doi:10.1155/2012/948390

[49] Glimcher LH, Kim KJ, Green I, Paul WE. Ia antigen bearing B cell tumor lines can present protein antigen and alloantigen in a major histocompatibility complex-restricted fashion to antigen-reactive T cells. J Exp Med. 1982; 155:445–459.

[50] Liu Q, Wang J, Boyd BJ. Peptide-based biosensors. Talanta. 2015; 136:114–127. doi: 10.1016/j.talanta.2014.12.020

[51] Palmieri C, Falcone C, Iaccino E, Tuccillo FM, Gaspari M, Trimboli F, De Laurentiis A, Luberto L, Pontoriero M, Pisano A, Vecchio E, Fierro O, Panico MR, Larobina M, Gargiulo S, Costa N, Dal Piaz F, Schiavone M, Arra C, Giudice A, Palma G, Barbieri A, Quinto I, Scala G. In vivo targeting and growth inhibition of the A20 murine B-cell lymphoma by an idiotype-specific peptide binder. Blood. 2010; 116:226–238. doi: 10.1182/blood-2009-11-253617

[52] De Angelis F, Pujia A, Falcone C, Iaccino E, Palmieri C, Liberale C, Mecarini F, Candeloro P, Luberto L, de Laurentiis A, Das G, Scala G, Di Fabrizio E. Water soluble nanoporous nanoparticle for in vivo targeted drug delivery and controlled release in B cells tumor context. Nanoscale. 2010; 2:2230–2236. doi:10.1039/c0nr00161a

[53] Migliaccio N, Palmieri C, Ruggiero I, Fiume G, Martucci NM, Scala I, Quinto I, Scala G, Lamberti A, Arcari P. B-cell receptor-guided delivery of peptide-siRNA complex for B-cell lymphoma therapy. Cancer Cell Int. 2015; 15:50–58. doi:10.1186/ s12935-015-0202-4

[54] Chen S, Liu L, Zhou J, Jiang S. Controlling antibody orientation on charged self-assembled monolayers. Langmuir. 2003; 19:2859–2864. doi:10.1021/la026498v

[55] Rea I, Oliviero G, Amato J, Borbone N, Piccialli G, Rendina I, De Stefano L. Direct synthesis of oligonucleotides on nanostructured silica multilayers. J Phys Chem C. 2010; 114:2617–2621. doi:10.1021/jp908440u

[56] Kim J, Seidler P, Wan LS, Fill C. Formation, structure, and reactivity of amino-termi-
nated organic films on silicon substrates. J Colloid Interface Sci. 2009; 329:114–119.
doi:10.1016/j.jcis.2008.09.031

[57] De Stefano L, Oliviero G, Amato J, Borbone N, Piccialli G, Mayol L, Rendina I,
Terracciano M, Rea I. Aminosilane functionalizations of mesoporous oxidized silicon
for oligonucleotide synthesis and detection. J R Soc Interface. 2013; 10:20130160. doi:
10.1098/rsif.2013.0160

[58] Martucci NM, Rea I, Ruggiero I, Terracciano M, De Stefano L, Migliaccio N, Palmieri
C, Scala G, Arcari P, Rendina I, Lamberti A. A new strategy for label-free detection of
lymphoma cancer cells. Biomed Opt Express. 2015; 6:1353–1362. doi:10.1364/BOE.
6.001353

[59] Martucci NM, Rea I, Ruggiero I, Terracciano M, De Stefano L, Migliaccio N, Dardano
P, Arcari P, Rendina I, Lamberti A. A silicon-based peptide biosensor for label-free
detection of cancer cells. Proceedings of SPIE 9506, Optical Sensors. 2015; 95061T. doi:
10.1117/12.2178137

6

Lumped-Element Modeling for Rapid Design and Simulation of Digital Centrifugal Microfluidic Systems

Mahdi Mohammadi, David J Kinahan and
Jens Ducrée

Abstract

Since the 1990s, centrifugal microfluidic platforms have evolved into a mature technology for the automation of bioanalytical assays in decentralized settings. These "Lab-on-a-Disc" (LoaD) systems have already implemented a range of laboratory unit operations (LUOs) such as sample loading, liquid transport, metering, aliquoting, routing, mixing, and washing. By assembling these LUOs in highly functional microfluidic networks, including sample preparation and detection, a sizable portfolio of common test formats such as general chemistry, immunoassays/ protein analysis, nucleic acid testing, and cell counting has been established. The availability of these bioanalytical assay types enables a broad range of applications in fields such as life-science research, biomedical point-of-care testing and veterinary diagnostics, as well as agrifood, environmental, infrastructural, and industrial monitoring.

Recently, a new method of the so-called "event-triggered" flow control has been developed which is independent of the spin rate. These valves, which function in a handshake mode as opposed to the typically batchwise liquid transfers on the "Lab-on-a-Disc" (LoaD) platform, assume a similarly pivotal role as relays and transistors in digital electronics, allowing conditional, logical (flow) control elements. This chapter will describe the modeling of this new generation of "digital" centrifugal microfluidic systems with low-dimensional, lumped-element simulations which have already been instrumental to the modern success story of modern microelectronics.

Keywords: lumped-element simulation, centrifugal microfluidics, lab-on-a-disc, event-triggered flow control, valving

1. Introduction

The centrifugal microfluidic platform has evolved into mature technology platform which has already proven to open significant market opportunity [1–5]. A large number of groups groups working on such LoaD systems in industry has already convincingly demonstrated the capability to integrate, automate, parallelize, and miniaturize a wide range of common bioanalytical test formats for detecting targets such as small molecules, proteins/antibodies, nucleic acids, and cells. Applications span from decentralized biomedical point-of-care diagnostics, veterinary medicine and agrifood, to the surveillance of the environment and infrastructures.

Based on the recently introduced, event-triggered flow control scheme [6, 7], highly functional microfluidic circuits can be assembled in a modular fashion from a limited set of LUOs to implement a broad repertoire of multi-step, multi-reagent bioassay protocols in a sample-to-answer fashion. Furthermore, it has been demonstrated that the chips could be progressively miniaturized to significantly enhance integration density (i.e., the number of assay steps and/ or tests per disc) and thus boost the overall cost efficiency and functionality of the LoaD platform.

Similar to integrated circuits in microelectronics, a microfluidic network can be modeled by lumped-element descriptors. Instead of a finely meshed 3-dimensional lattice, the Lab-on-a-Disc systems are described by a low number of parameters such as pressure head (voltage source), flow resistance (electric resistance), and compressibility (capacitance). This reduced-dimension method can be utilized for fast design and simulation of microfluidic systems that are composed of a library of functional units.

After the introduction (Section 1), the hydrodynamic principles of centrifugal microfluidics are presented (Section 2) before outlining digital flow control schemes (Section 3). Next, the concept of lumped-element simulation in event-triggered centrifugal microfluidic networks is developed (Section 4).

2. Centrifugal hydrodynamics

The rotationally controlled microfluidic "Lab-on-a-Disc" platform is based on (the volume density of) the centrifugal force

$$f_\omega = \rho r \omega^2 \tag{1}$$

the Euler force

$$f_E = \rho r . d\omega / dt \tag{2}$$

and the Coriolis force [1]

$$f_C = 2\rho\omega v \tag{3}$$

where ρ is a fluid density on a rotating platform, $\omega = 2\pi v$ the angular velocity with the frequency of rotation v, r is a distance from a central axis, and v represents the speed of flow. All forces act in the plane of the disc and scale with the angular velocity ω directly impacts these three forces.

The centrifugal force (Eq. (1)) translates into an equivalent centrifugal pressure

$$\Delta p_\omega = \rho\Delta r\bar{r}\omega^2 \tag{4}$$

and an average flow velocity

$$V = \frac{D_h^2 \rho\Delta r\omega^2\bar{r}}{32\mu L} \tag{5}$$

of the liquid in the channel [5] featuring the hydraulic diameter $D_h = 4A/P$ with the A, P, and L, its cross-sectional area A, wetted perimeter of the channel P, and length L. The fluid viscosity is denoted by μ, the mean radial position by $\bar{r} = (r_2 + r_1)/2$, and the radial length by $\Delta r = (r_2 - r_1)$.

Air pockets, which often arise accidentally or strategically during priming, can be compressed by the hydrostatic pressure head of a liquid column (4) in a more central position. This centrifugally induced pressure compresses the enclosed gas volume

$$P_c = P_0 \frac{1}{1 - \Delta V/V} \tag{6}$$

according to Boyle's law [6] where P_c represents the pressure of the gas in the pneumatic chamber, V is the total volume of the pneumatic chamber, and ΔV denotes a reduction of gas volume due to filling of liquid in the pneumatic chamber.

3. Digital flow control schemes

Flow control is instrumental for orchestrating sequential liquid handling on the LoaD platforms where all volumes are subjected to the same centrifugal field (Eq. (1)). Such flow control can be categorized into rotationally actuated and instrument-supported schemes.

Instrument-supported valves involve some stationary modules (other than the platform innate spindle motor). To switch a valve, these "lab-frame" elements interact with the disc cartridge, either at rest or during spinning. The actuation can be powered by pneumatic pressure sources [8–9], heating of phase-change materials [10–13], or even varying the chip orientation with respect to the radial direction [14–16]. While these may provide enhanced and more flexible control, these active valving mechanisms typically involve additional instrumentation, maintenance, cost, and susceptibility to failure.

Rotationally actuated valves are far more common and are considered more suitable for deployment of inexpensive point-of-use applications. Through varying the rotationally induced fields relative to the statically defined forces such as interfacial or membrane tension, the force equilibrium at a fluid element can be unbalanced. Such static forces can be implemented by capillary action [17–21], dissolvable films (DFs) [22- 23], burstable foils [24], elastomeric membranes [25], dead-end pneumatic chambers [26], siphons [27]–[29], and pneumatically enhanced centrifugo-pneumatic siphons (CPSVs) [30–33].

In particular the popular, rotationally actuated capillary "burst" valves are strongly dependent on physicochemical properties such as geometry, surface roughness, and contact angle; hence, valve performance is intimately linked to manufacturing fidelity. The often rather poor reproducibility and stability of these effects translate into a significant "smearing" of the burst frequencies. For serial flow control which is common in bioanalytical protocols, rather wide, non-overlapping bands of the spin rate have to be reserved for each assay step. As the maximum spin rate is practically limited by the motor power and safety, this imposes a practical limit on the number of sequential LUOs which can be rotationally controlled by a spindle motor.

Event-triggered valving circumvents this restriction [6– 7]. Here, the arrival of liquid at defined locations on the disc coordinates a sequential opening of valves; valve actuation is thus decoupled from changes in the spin rate and support instrumentation. So far, event-triggered valving has been based on dissolvable film (DF) membranes [22], [23], [34], [35] and, in function, can be described akin to an electrical relay. The architecture of the disc determines the order of valve actuation, while the timing is controlled by the dissolution of these membranes. It has been shown that event-triggered schemes can also implement logical flow control elements such as AND and OR conditions [6], thus enabling a modular system design similar to electronics. Developing the lumped-element tool for the simulation, digital centrifugal microfluidic systems can generate a broad scope of applications, thus mitigating development risks, upfront investment, and time to market.

The basic event-triggered valve is composed of a pneumatic chamber sealed by the restrained liquid and two dissolvable films called the load film (LF) and the control film (CF). The geometry of the pneumatic chamber is designed so that, at the spin rates typical for the centrifugal platform, the restrained liquid cannot be pumped into contact with the LF or CF by compressing the trapped air within the pneumatic chamber. Similarly, the section of the chamber connecting the LF and CF extends radially inward of the restrained liquid. When the CF is wetted and dissolved by an ancillary liquid, the pneumatic chamber is vented so the main liquid contacts and thus opens the LF.

However, the connecting channel between the LF and CF acts as a geometric barrier which prevents the liquid escaping through the disrupted CF. Thus, in this configuration, the CF acts analogous to the control line of an electrical relay and the LF to the load line. This basic configuration can then be arranged into a complex fluid network where valves sequentially cascaded; the flow released from the first valve triggers the subsequent "ancillary liquid." Importantly, the interval between valve actuations is governed by the aggregate time of membrane dissolution and liquid transfer.

Alongside the basic configuration (**Figure 1**), the conditions of valve actuation can be altered by changing the arrangement of the CF. For example, locating the CF such that it can only be wetted when two or more upstream "ancillary liquid' volumes have been released (Figure 1) establishes a Boolean AND condition. Similarly, by designing a valve with two CFs where wetting one or the other will trigger the valve, we create a Boolean OR condition (**Figure 2**). Finally, locating two CFs in close proximity so they can be reached by a single ancillary liquid can simultaneously open two pneumatically connected valves and thus can represent parallel valve actuation (**Figure 3**).

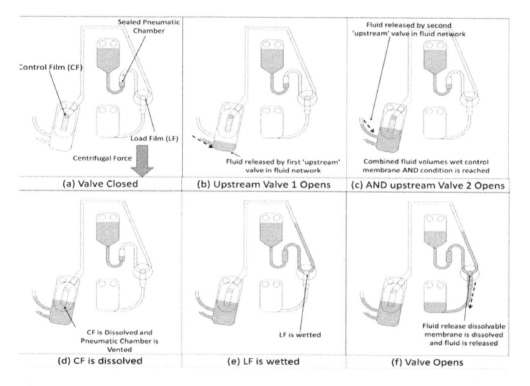

Figure 1. Schematic demonstrating the basic event-triggered configuration and also showing the Boolean AND release mechanism (a) Valve closed, (b) upstream valve 1 opens (c) AND upstream valve 2 opens, (d) CF is dissolved, (e) LF is wetted, (f) Valve opens.

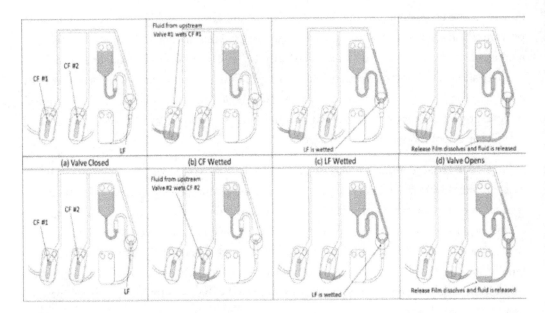

Figure 2. Schematic illustrating the OR conditional release mechanism. The top pane shows the valve actuation triggered by liquid movement to one chamber and the lower pane shows the valve actuation triggered. (a) Valve closed, (b) CF wetted, (c) LF wetted, (d) Valve opens.

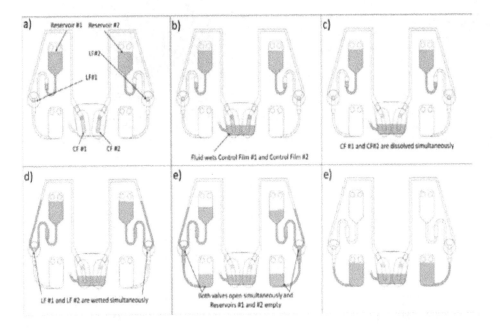

Figure 3. Schematic illustrating the OR conditional release mechanism. The top pane shows the valve actuation triggered by liquid movement to one chamber and the lower pane shows the valve actuation triggered. (a) Valve closed, (b) CF wetted, (c) LF wetted, (d) Valve opens.

4. The concept of lumped-element simulation in digital, event-triggered centrifugal microfluidic networks

The rapid evolution of microelectronics (following Moore's law) has been leveraged by the trinity of miniaturization, fabrication, and, last but not least, large-scale system integration (LSI). The breathtaking progress within these tightly intertwined factors has tremendously reduced production costs and seminally enhanced system performance. This is clearly visible when looking back over the last decades when microelectronic devices took the road from very clumsy, maintenance-intensive, multi-million dollar machines sparsely scattered around the globe to the sleek, ubiquitous, and quite affordable digital gadgets people even carry in their pockets. The unprecedented commercial success story of microelectronics has been enabled by seminal advances in microfabrication as well as the capability to generate complex functional architectures from a limited set of base modules such as capacitors and transistors. These simple modules are composed into sophisticated functional networks by lumped-element model software. We have developed a new type of "digital" LoaD platform which follows a similar design paradigm to implement different types of bioanalytical tests, e.g., for small molecules, proteins, antibodies, DNA, and cells [6].

Over the past decades, simulation has a key role in developing new products. The common simulation methods are FEA (finite element analysis), CFD (computational fluid dynamic), and MBS (multi-body systems). In principle, these mesh-based simulation methods are very accurate. Nevertheless, these numerical tools display serious limitations, for instance, that they tend to be very time-consuming; in particular for more complex networks, also the grid size and proper boundary conditions impact the result (mesh dependency). Therefore, simplified geometries are required for keeping computation times and common convergence issues at bay; lumped-element simulation was proposed to simplify analysis based on electric circuit elements; this method is quite fast and fit for swift parameter optimization; in addition, these methods could simulate serial and parallel multi-element architectures [36].

The centrifugal flow control elements and their combination of complex microfluidic circuitry translate into equivalent, lumped-element descriptors. Each lumped element exhibits certain free parameters, for instance, corresponding to resistances or capacitances. In microfluidics, these parameters typically relate to geometries, e.g., the channel cross section, as well as hydrodynamic and mechanical properties such as the viscosity and compressibility of the fluids and the flexibility of the ducts. Lumped-element analogies for the different environments are listed in **Table 1**.

	Effort (e)	Flow (f)	Inertance	Capacitance	Resistance	Displacement (q)	Node law	Mesh law
Electricity,...	Voltage (V)	Current (I)	Inductor (L)	Capacitor (C)	Resistor (R)	Charge (Q)	KCL	KVL
Fluidic	Pressure (P)	Flow (q)	Inertance (M)	Fluid capacitance	Flow resistance	Volume (V)	Mass conservation	Pressure is relative

	Effort (e)	Flow (f)	Inertance	Capacitance (C)	Resistance (R)	Displacement (q)	Node law	Mesh law
Mechanics	Force (F)	Velocity (V)	Mass (m)	Spring (K)	Damper (b)	Displacement (x)	Continuity of space	Newton's 2nd law
Thermal	Temp. diff (Δ T)	Heat flow	–	Heat capacity (mcp)	Thermal resistance (R)	Heat (Q)	Heat energy conservation	Temperature is relative

Table 1. Physical lumped-element analogies in different environments.

For a given microfluidic network and spin rate protocol, the lumped-element simulation of microfluidic systems allows to calculate pressure distribution, flow rate, and timing. Parallel simulation and parameter sweep for efficient design generation of microfluidic systems represent further advantages of lumped-element simulation. In addition, its computational simplicity and fast convergence mean it can also be applied to "real-time" active control of microfluidic processes. Utilizing this real-time graphical simulation to monitor filling level and aliquoting timing along the LUOs in multi-step, multi-reagent bioassay protocols will constitute an important milestone because it would allow the evaluation of the functional operation of the LoaD device without any further fabrication and experimental processes.

This lumped-element simulation in different environments is illustrated by the equivalent electric circuit elements comprising a resistor, a capacitor, and a diode, and the required relations for lumped-element simulation are presented in the following:

4.1. Kirchhoff's current law (KCL) [37]

The material balance equation, flow-in equal flow-out at any given node in the microfluidic network.

$$\sum_i Ii = 0 \xrightarrow{\text{At each nodes}} \sum_i qi = 0 \rightarrow q_{in(1)} + q_{in(2)} + \ldots = q_{out(1)} + q_{out(2)} + \ldots \tag{7}$$

4.2. Kirchhoff's voltage law (KVL) [31]

The sum of pressure differences around a microfluidic loop must be zero.

$$\sum_i Vi = 0 \xrightarrow{\text{In closed loop}} \sum_i Pi = 0 \tag{8}$$

4.3. Capacitance

Increasing charge storage results in increasing voltage in an electrical capacitor and increasing fluid leads to increase pressure in the reservoir (fluid capacitator).

$$V_L = C.P \tag{9}$$

The force at the bottom of storage due to the weight is mg = $\rho V_L g$ which constant earth gravitational replaces by artificial gravity field $g = \bar{r}\omega^2$ in the centrifugal microfluidic system.

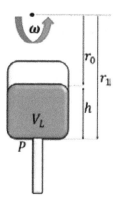

$$P = \frac{\rho V_L g}{A} = \rho g h \rightarrow P = \frac{\rho V_L \bar{r} \omega^2}{A} = \rho \bar{r} \omega^2 (r_1 - r_0)$$

$$V_L = \left[\frac{A}{\rho \bar{r} \omega^2} \right] P \tag{10}$$

The fluid capacitance in centrifugal system is $C = \left[\frac{A}{\rho \bar{r} \omega^2} \right]$.

4.4. Flow resistance

The flow resistance can be considered Ohm's law $\Delta V = IR$.

$$\Delta P = qR \tag{11}$$

Flow resistance of rectangular microchannel can be calculated using the following Fourier series [37].

$$R_h = \frac{12 \eta L}{\left(1 - \frac{h}{w} \left(\frac{192}{\pi^5} \sum_{n=1,2,3}^{\infty} \frac{1}{n^5} \tan h \left(\frac{n \pi w}{2h} \right) \right) \right) \omega h^3} \tag{12}$$

$$P = I\dot{q}$$

Also, flow resistance by rectangular cross section for $h/w \ll 1$ can be approximated [31]:

$$R_h = \frac{12\eta L}{\omega h^3} \tag{13}$$

where w is width and h height of the channel.

4.5. Inertance

Newton's second law that is called the linear momentum relation of fluid flow in the channel is [38]:

$$\sum \vec{F} = \frac{d\left(m\vec{V}\right)}{dt} \tag{14}$$

$$A\left(P_1 - P_2\right) = m\dot{v} = \rho L A \dot{v} \tag{15}$$

$$\Delta P = \frac{\rho L}{A}\dot{q} \tag{16}$$

This relation is similar to inductor equation $\Delta V = L\frac{di}{dt}$ and we could write where $I = \rho L/A$ and p represents pressure difference.

4.6. Application example

In this work, we consider a single design which allows us to demonstrate how our lumped-element approach can be applied to "digital" centrifugal flow control. Therefore, we model a liquid handling protocol similar to that used by Nwankire et al. [35] to implement a nitrite/nitrate panel for whole blood monitoring. To implement their assay, Nwankire et al. used DF burst valves which were designed to open in sequence with increasing spin rate of the disc. We present a lumped-element model to simulate the centrifugo-pneumatic chambers which are the key enabling technology of the DF burst valves; a good understanding of these chambers is also critical to the implementation of our event-triggered valving architecture [22]. The schematic view of the design is shown in **Figure 4**. This design shows three reservoirs, labeled A through C and three pneumatic chambers which are sealed using DF burst valves. The DFs are arranged to burst at a rotational frequency greater than 20 Hz and less than 40 Hz. These reservoirs feed a mixing chamber which is further sealed by two DFs which dissolve on contact with the liquid. Upon dissolution, an open path into two overflow reservoirs is

provided.

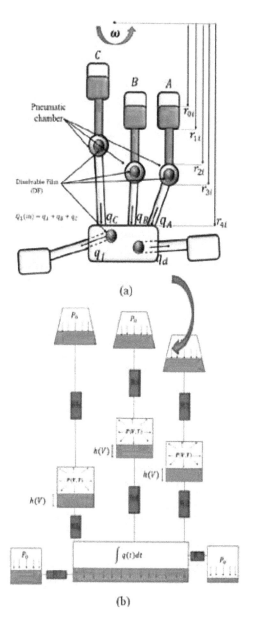

Figure 4. (a)The design of centrifugal microfluidic platform (b) Schematic view of Lumped element network.

We model the system in four different conditions to demonstrate a parallel simulation defined by different spin rates. These conditions share a spin profile (**Figure 5a**) which involves a rapid acceleration to a maximum frequency, followed by rapid mixing, stopping the disc, and then fast acceleration back to the maximum frequency. These spin protocols are identical except for their magnitude; they have maximum spin rates of 20 Hz, 40 Hz, 60 Hz, and 80 Hz.

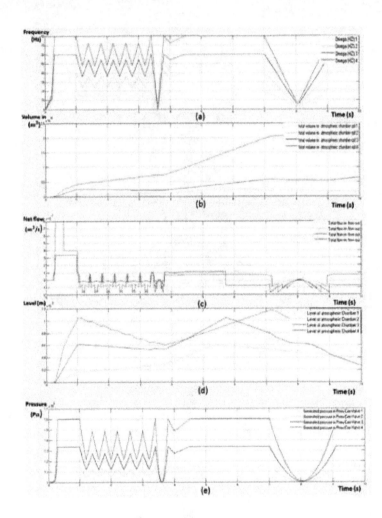

Figure 5. Lumped-element simulation graph for main atmospheric reservoir. (a) Angular frequency profile vs the time. (b) Total inflow vs time. (c) Total net flow (flow-in-flow-out) vs time. (d) Filling level of the chamber due to the ingress of liquid into the reservoir. (e) Pressure generated in the pneumatic and centrifugal valve (A) due to angular velocity.

To demonstrate the wide capability of this lumped-element model to predict on disc performance, a number of parameters are shown in **Figure 5** which have been calculated using the simulation software based on a number of defined boundaries and initial conditions. These parameters are the volume flow into the mixing chamber assuming no out volume flow through the two exits (**Figure 5b**); the net flow into the mixing chamber, assuming outflow through the exists (**Figure 5c**); and the liquid level in the mixing chamber, assuming outflow through the exits (**Figure 5d**). Finally, Figure 5e shows the predicted pressure, during the spin profile, in each DF burst valve with the assumption the DF does not dissolve.

Based on the lumped-element analysis, the critical burst frequencies of the DFs are between 25 and 30 Hz. Therefore, in **Figure 5b, c**, and d, it is predicted that, for the 20 Hz test condition, the DFs do not dissolve and so there is no liquid flow. As stated above, Figure 5b shows the total volume entering into the main chamber; this is defined as $V_{in} = V_A + V_B + V_C$. Similarly, the net flow rate in and out of the mixing chamber is shown in **Figure 5c** and is defined by $q_{net} = q_A + q_B + q_C - q_d - q_f$. In turn, and most importantly, the liquid level in the main chamber can also be predicted in Figure 5d; this information is important as it can be used to inform incubation times and washing protocols which are critical for Lab-on-a-Disc applications.

Finally, in **Figure 5e**, the pressurization of the centrifugo-pneumatic valves (Valve A) is presented. Here, the increased centrifugal force pushes the liquids from the main reservoir (Reservoir A) and into the dead-end pneumatic chamber which is sealed by a DF. The fluid flow is stopped in the pneumatic chamber by a pocket of entrapped air which pushes back against the centrifugally generated hydrostatic pressure; this equilibrium condition is reached when the centrifugal pressure head, described previously in Eq. (4), balances with the pressure of the trapped gas, defined by Boyle's law in Eq. (6). In the real case, the DF membrane in the pneumatic chamber is dissolved (valve opening) beyond the critical burst frequency when the liquid ingress is sufficient to contact the film. Then, the liquid flows are directed into the main (downstream) chamber.

Over the past three decades, a special breed of microfluidic systems is based on centrifugal liquid handling for a wide spectrum of applications in biomedical point-of-care diagnostics and the life sciences. Recently, event-triggered flow control was introduced on these LoaD platforms to implement logical flow control which functions akin to digital microelectronics [33]. Similar to the difference between an old-fashioned office mainframe and a modern smartphone, these breakthroughs may provide an unprecedented level of system integration and automation which is needed to eventually implement complex, highly functional networks representing a repertoire of bioanalytical assays on a user-friendly, cost-efficient, portable, and still high-performance microfluidic point-of-use "gadget." We presented an advanced lumped-element approach for the fast-generation and robust simulation for event-triggered centrifugal microfluidic networks.

Author details

Mahdi Mohammadi, David J Kinahan and Jens Ducrée*

*Address all correspondence to: jens.ducree@dcu.ie

School of Physical Sciences, National Centre for Sensor Research, Dublin City University (DCU), Dublin, Ireland

References

[1] J. Ducrée, S. Haeberle, S. Lutz, S. Pausch, F. Von Stetten, and R. Zengerle, "The centrifugal microfluidic Bio-Disk platform," *J. Micromechanics Microengineering*, vol. 17, no. 7, pp. S103–S115, 2007.

[2] S. Smith, D. Mager, A. Perebikovsky, E. Shamloo, D. Kinahan, R. Mishra, S. Torres Delgado, H. Kido, S. Saha, J. Ducrée, M. Madou, K. Land, and J. Korvink, "CD-Based Microfluidics for Primary Care in Extreme Point-of-Care Settings," *Micromachines*, vol. 7, no. 2, p. 22, 2016.

[3] O. Strohmeier, M. Keller, F. Schwemmer, S. Zehnle, D. Mark, F. von Stetten, R. Zengerle, and N. Paust, "Centrifugal microfluidic platforms: advanced unit operations and applications.," *Chem. Soc. Rev.*, vol. 44, pp. 6187–6229, 2015.

[4] R. Gorkin, J. Park, J. Siegrist, M. Amasia, B. S. Lee, J.-M. Park, J. Kim, H. Kim, M. Madou, and Y.-K. Cho, "Centrifugal microfluidics for biomedical applications.," *Lab Chip*, vol. 10, pp. 1758–1773, 2010.

[5] M. Madou, J. Zoval, G. Jia, H. Kido, J. Kim, and N. Kim, "Lab on a Cd," *Annu. Rev. Biomed. Eng.*, vol. 8, no. 1, pp. 601–628, 2006.

[6] D. J. Kinahan, S. M. Kearney, N. Dimov, M. T. Glynn, and J. Ducrée, "Event-triggered logical flow control for comprehensive process integration of multi-step assays on centrifugal microfluidic platforms," *Lab Chip, vol.* 14, no. 13, pp. 2249–2258, 2014.

[7] D. J. Kinahan, S. M. Kearney, O. P. Faneuil, M. T. Glynn, N. Dimov, and J. Ducrée, "Paper imbibition for timing of multi-step liquid handling protocols on event-triggered centrifugal microfluidic lab-on-a-disc platforms," *RSC Adv.*, vol. 5, no. 3, pp. 1818"1826, 2015.

[8] M. Kong and E. Salin, "Pnenumatic flow switching on centrifugal microfluidic platforms in motion," *Anal. Chem.*, vol. 83, pp. 1148–1151, 2011.

[9] L. Clime, D. Brassard, M. Geissler, and T. Veres, "Active pneumatic control of centrifugal microfluidic flows for lab-on-a-chip applications.," *Lab Chip*, vol. 15, no. 11, pp. 2400–11, 2015.

[10] J. L. Garcia-Cordero, D. Kurzbuch, F. Benito-Lopez, D. Diamond, L. P. Lee, and A. J. Ricco, "Optically addressable single-use microfluidic valves by laser printer lithography," *Lab Chip*, vol. 10, no. 20, p. 2680, 2010.

[11] B. S. Lee, Y. U. Lee, H.-S. H. Kim, T.-H. Kim, J. Park, J.-G. Lee, J. Kim, H.-S. H. Kim, W. G. Lee, and Y.-K. Cho, "Fully integrated lab-on-a-disc for simultaneous analysis of biochemistry and immunoassay from whole blood.," *Lab Chip*, vol. 11, no. 1, pp. 70"78, 2011.

[12] K. Abi-Samra, R. Hanson, M. Madou, and R. a Gorkin, "Infrared controlled waxes for liquid handling and storage on a CD-microfluidic platform.," *Lab Chip*, vol. 11, no. 4, pp. 723–726, 2011.

[13] W. Al-Faqheri, F. Ibrahim, T. H. G. Thio, J. Moebius, K. Joseph, H. Arof, and M. Madou, "Vacuum/Compression Valving (VCV) Using Parrafin-Wax on a Centrifugal Microfluidic CD Platform," *PLoS One*, vol. 8, no. 3, pp. 2–10, 2013.

[14] T. Kawai, N. Naruishi, H. Nagai, Y. Tanaka, Y. Hagihara, and Y. Yoshida, "Rotatable reagent cartridge for high-performance microvalve system on a centrifugal microfluidic device," *Anal. Chem.*, vol. 85, no. 14, pp. 6587–6592, 2013.

[15] M. Geissler, L. Clime, X. D. Hoa, K. J. Morton, H. Hébert, L. Poncelet, M. Mounier, M. Deschênes, M. E. Gauthier, G. Huszczynski, N. Corneau, B. W. Blais, and T. Veres, "Microfluidic Integration of a Cloth-Based Hybridization Array System (CHAS) for Rapid, Colorimetric Detection of Enterohemorrhagic Escherichia coli (EHEC) Using an Articulated, Centrifugal Platform," *Anal. Chem.*, vol. 87, no. 20, pp. 10565–10572, 2015.

[16] B. Miao, N. Peng, L. Li, Z. Li, F. Hu, Z. Zhang, and C. Wang, "Centrifugal Microfluidic System for Nucleic Acid Ampli?cation and Detection," *Sensors*, vol. 15, no. 11, pp. 27954–27968, 2015.

[17] T. Li, L. Zhang, K. M. Leung, and J. Yang, "Out-of-plane microvalves for whole blood separation on lab-on-a-CD," *J. Micromechanics Microengineering*, vol. 20, no. 10, p. 105024, 2010.

[18] S. Haeberle, T. Brenner, R. Zengerle, and J. Ducrée, "Centrifugal extraction of plasma from whole blood on a rotating disk.," *Lab Chip*, vol. 6, no. 6, pp. 776–781, 2006.

[19] T. H. G. Thio, S. Soroori, F. Ibrahim, W. Al-Faqheri, N. Soin, L. Kulinsky, and M. Madou, "Theoretical development and critical analysis of burst frequency equations for passive valves on centrifugal microfluidic platforms.," *Med. Biol. Eng. Comput.*, vol. 51, no. 5, pp. 525–35, 2013.

[20] J. L. Moore, A. McCuiston, I. Mittendorf, R. Ottway, and R. D. Johnson, "Behavior of capillary valves in centrifugal microfluidic devices prepared by three-dimensional printing," *Microfluid. Nanofluidics*, vol. 10, no. 4, pp. 877–888, 2011.

[21] J. M. Chen, P. C. Huang, and M. G. Lin, "Analysis and experiment of capillary valves for microfluidics on a rotating disk," *Microfluid. Nanofluidics*, vol. 4, no. 5, pp. 427–437, 2008.

[22] R. Gorkin, C. E. Nwankire, J. Gaughran, X. Zhang, G. G. Donohoe, M. Rook, R. O'Kennedy, and J. Ducree, "Centrifugo-pneumatic valving utilizing dissolvable films," *Lab a Chip - Miniaturisation Chem. Biol.*, vol. 12, no. 16, pp. 2894–2902, 2012.

[23] C. E. Nwankire, M. Czugala, R. Burger, K. J. Fraser, T. M. Connell, T. Glennon, B. E. Onwuliri, I. E. Nduaguibe, D. Diamond, and J. Ducrée, "A portable centrifugal analyser for liver function screening," *Biosens. Bioelectron.*, vol. 56, pp. 352–358, 2014.

[24] T. van Oordt, Y. Barb, J. Smetana, R. Zengerle, and F. von Stetten, "Miniature stick-packaging – an industrial technology for pre-storage and release of reagents in lab-on-a-chip systems," *Lab Chip*, vol. 13, no. 15, p. 2888, 2013.

[25] H. Hwang, H.-H. Kim, and Y.-K. Cho, "Elastomeric membrane valves in a disc.," *Lab Chip*, vol. 11, no. 8, pp. 1434–1436, 2011.

[26] D. Mark, P. Weber, S. Lutz, M. Focke, R. Zengerle, and F. Von Stetten, "Aliquoting on the centrifugal microfluidic platform based on centrifugo-pneumatic valves," *Microfluid. Nanofluidics*, vol. 10, no. 6, pp. 1279–1288, 2011.

[27] C. E. Nwankire, G. G. Donohoe, X. Zhang, J. Siegrist, M. Somers, D. Kurzbuch, R. Monaghan, M. Kitsara, R. Burger, S. Hearty, J. Murrell, C. Martin, M. Rook, L. Barrett, S. Daniels, C. McDonagh, R. O'Kennedy, and J. Ducrée, "At-line bioprocess monitoring by immunoassay with rotationally controlled serial siphoning and integrated supercritical angle fluorescence optics.," *Anal. Chim. Acta*, vol. 781, no. 4, pp. 54–62, 2013.

[28] J. Siegrist, R. Gorkin, L. Clime, E. Roy, R. Peytavi, H. Kido, M. Bergeron, T. Veres, and M. Madou, "Serial siphon valving for centrifugal microfluidic platforms," *Microfluid. Nanofluidics*, vol. 9, no. 1, pp. 55–63, 2010.

[29] M. Kitsara, C. E. Nwankire, L. Walsh, G. Hughes, M. Somers, D. Kurzbuch, X Zhang, G. G. Donohoe, R. O'Kennedy, and J. Ducrée, "Spin coating of hydrophilic polymeric films for enhanced centrifugal flow control by serial siphoning," *Microfluid. Nanofluidics*, vol. 16, no. 4, pp. 691–699, 2014.

[30] M. M. Aeinehvand, F. Ibrahim, S. W. harun, W. Al-Faqheri, T. H. G. Thio, A. Kazemzadeh, and M. Madou, "Latex micro-balloon pumping in centrifugal microfluidic platforms," *Lab Chip*, vol. 14, no. 5, p. 988, 2014.

[31] F. Schwemmer, S. Zehnle, D. Mark, F. von Stetten, R. Zengerle, and N. Paust, "A microfluidic timer for timed valving and pumping in centrifugal microfluidics," *Lab Chip*, vol. 15, no. 6, pp. 1545–1553, 2015.

[32] N. Godino, R. Gorkin, A. V Linares, R. Burger, and J. Ducrée, "Comprehensive integration of homogeneous bioassays via centrifugo-pneumatic cascading.," *Lab Chip*, vol. 13, no. 4, pp. 685–94, 2013.

[33] R. Gorkin, L. Clime, M. Madou, and H. Kido, "Pneumatic pumping in centrifugal microfluidic platforms," *Microfluid. Nanofluidics*, vol. 9, no. 2–3, pp. 541–549, 2010.

[34] C. E. Nwankire, A. Venkatanarayanan, T. Glennon, T. E. Keyes, R. J. Forster, and J. Ducrée, "Label-free impedance detection of cancer cells from whole blood on an integrated centrifugal microfluidic platform.," *Biosens. Bioelectron.*, vol. 68C, no. November, pp. 382–389, 2014.

[35] C. Nwankire, D.-S. Chan, J. Gaughran, R. Burger, R. Gorkin, and J. Ducrée, "Fluidic Automation of Nitrate and Nitrite Bioassays in Whole Blood by Dissolvable-Film Based Centrifugo-Pneumatic Actuation," *Sensors*, vol. 13, no. 9, pp. 11336–11349, 2013.

[36] R. Z. and P. K. C. Litterst, W. Streule, "Simulation Toolkit for Micro Fluidic Pumps Using Lumped Element Model," vol. 1, pp. 736–739, 2005.

[37] K. W. Oh, K. Lee, B. Ahn, and E. P. Furlani, "Design of pressure-driven microfluidic networks using electric circuit analogy," *Lab Chip*, vol. 12, no. 3, pp. 515–545, 2012.

[38] F. White, Fluid Mechanics, McGraw-Hill, New York, NY, USA: 3–225. 1999.

7

CMOS Circuits and Systems for Lab-on-a-Chip Applications

Yehya H. Ghallab and Yehea Ismail

Abstract

Complementary metal oxide semiconductor (CMOS) technology allows the functional integration of sensors, signal conditioning, processing circuits and development of fully electronic integrated lab-on-a-chip. On the other hand, lab-on-a-chip is a technology which changed the traditional way by which biological samples are inspected and tested in laboratories. A lab-on-a-chip consists of four main parts: sensing, actuation, readout circuit and microfluidic chamber. Lab-on-a-chip gives the promise of many advantages including better and improved performance, reliability, portability and cost reduction. This chapter reviews the currently used lab-on-a-chips based on CMOS technology. Also, this chapter presents and discusses the features of the existing CMOS based lab-on-a-chips and their applications at the cell level.

Keywords: lab-on-a-chip, biochips, biological cells, CMOS technology, integrated sensors

1. Introduction

Integration and miniaturization are the two main factors in the device engineering research, and these two factors are also the main features of the lab-on-a-chip technology. The main contributor for this major trend is the fast advances of integrated circuit (IC) technology. As a result of miniaturization, portable and cheaper equipment replaced the traditional bulky and expensive equipment.

In the field of cell biology, the use of micro- and nano-fabricated lab-on-a-chip using active substrates and integrated sensors [1] is still at experimental stage. In order to replace traditional optical inspection techniques, these techniques have been tested for the identification of objects at microscopic level to envision several applications [2].

Many research groups believe that lab-on-a-chip technology will be the major contributor to novel diagnostic tools. The target of these research groups is to have lab-on-a-chip systems that will allow healthcare suppliers and workers in poorly equipped clinics to do different tests which need more sophisticated equipment in a simpler way, for example, nucleic acid assays and immunoassays tests [3]. Thus, lab-on-a-chip innovation is a vital part of the endeavors to upgrade and enhance worldwide healthcare system, especially through the advancement of clinics and laboratories' testing techniques [4].

One dynamic area of lab-on-a-chip research includes approaches to analyze and oversee HIV infections. In view of UNAIDS 2014 report [5], more than 40 million individuals are tainted with HIV around the world; 13 million of these individuals get hostile to retroviral treatment. Around 55% of individuals with HIV have never been tested for the infection [5].

Measuring the quantity of CD4+ T lymphocytes in a man's blood is an exact and right approach to figure out whether a man has HIV and to track the advancement of a HIV contamination. Right now, flow cytometry is the brilliant standard for acquiring CD4 checks; yet, flow cytometry is a confounded method that is not accessible in most developing regions since it requires prepared professionals and costly instruments [6, 7].

The vicinity of the complementary metal oxide semiconductor (CMOS) technology allows the integration of sensors, amplifiers, filters and other circuitry on a single chip. Also, CMOS technology leads to a totally electronic integrated lab-on-a-chip utilizing a homogenous technology [8, 9]. CMOS is an innovation used in building integrated circuits. It is used in microcontrollers, static RAM, registers, microchips and other digital circuits. CMOS technology is utilized also for a wide assortment of analog circuits, for example, image sensors, amplifiers, analog to digital converters, and transceivers for communication modules. Low static power consumption and high noise immunity are two advantages of CMOS devices [10].

Numerous specialists and research teams have begun to utilize the CMOS technology in lab-on-a-chip applications [11, 12]. These applications incorporate, yet not restricted to, identification of neurons' activities, microorganism discovery and portrayal, electric field imaging, flow cytometry and polymerase chain reaction (PCR) applications [13–17].

This chapter covers and presents the state of the art in CMOS circuits and systems for lab-on-a-chip applications. It summarizes and reviews different circuits and systems of CMOS-based lab-on-a-chip technology at the cell level. These circuits and systems include: polymerase chain reaction (PCR), microorganism detection and characterization, neuronal activity detection, flow cytometry and electric field imaging applications.

2. Lab-on-a-chip based on CMOS technology

2.1. Polymerase chain reaction (PCR)

CMOS-based lab-on-a-chip has numerous applications in the science and biomedical fields. It is likewise utilized as a part of numerous different fields, for example, environmental applications. In medicine, lab-on-a-chip can be utilized as a part of real-time polymerase chain reaction (PCR) applications to identify bacteria, viruses and cancers [18, 19]. PCR is a biochemical technology in molecular biology to duplicate one or few DNA pieces, creating thousands to a huge number of duplicates of a specific DNA sequence [20].

The incredible preference about PCR is in the snappy and specific enhancements of target qualities through a cyclic and enzyme-catalyzed [21]. PCR procedures have been connected to mutation analysis and DNA [18, 19, 21]. As of late, Norian et al. [22] displayed an ongoing PCR lab-on-a-chip that was implemented using 0.35 μm CMOS technology. It performs electrowetting droplet-based transport, reagent heating, temperature sensing and integrated fluorescence measurements.

Figure 1 demonstrates the 4 × 4mm chip. It has a 7 × 8 array of 200 × 200 μm electrodes [22]. Using electrowetting-on-dielectric transport technique, each droplet can be dislodged [23]. The three temperature zones on the top of the chip are managed by a heater and temperature sensors. Based on this PCR lab-on-a-chip, a droplet can be allocated at a thermal cycled area on the chip. Also, a thermal equilibrium can be obtained in a 500 ms at most, if a droplet's volume is under 1nL. Likewise, fluorescent measurement using integrated Geiger-mode single-photon avalanche photodiodes (SPADs) allows for sensitive fluorescent detection. **Figures 2** and **3** show the on-chip thermal regulation and PCR thermal cycling on chip, respectively.

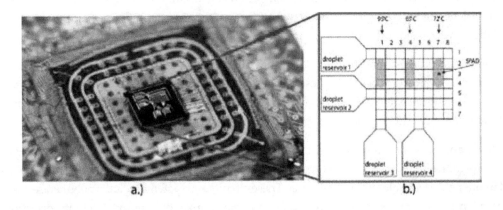

Figure 1. PCR Chip [22].

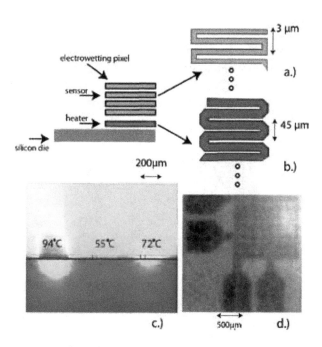

Figure 2. On-chip thermal regulation. (a) Layout of metal-resistor temperature sensors; (b) layout of polysilicon heaters; (c) cross-sectional COMSOL simulation; (d) IR image displays distinct temperature region generated by a single polysilicon heater [22].

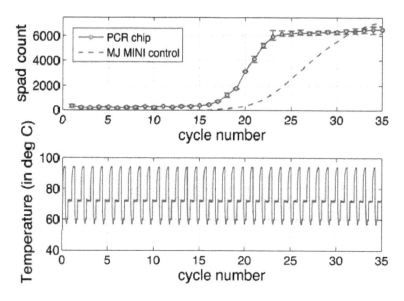

Figure 3. PCR thermal cycling on chip. (a) Fluorescence measurements for *Staphylococcus aureus* target; (b) on-chip temperature profile [22].

2.2. Dielectrophoresis (DEP)

Lab-on-a-chip based on CMOS technology is utilized as a part of biochemical tests, where it is utilized to quantify the presence of a target element which can be a biochemical substance or a medication or a cell in a sample [24]. Diverse range of uses is the dielectrophoresis applications where CMOS-based lab-on a-chip is utilized to recognize microbes, bacteria and cancer cells [14].

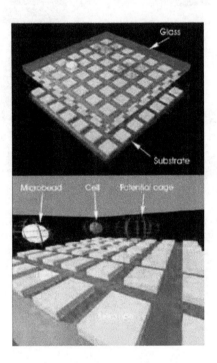

Figure 4. The lab-on-a-chip architecture for cells manipulation and sensing [25].

Manaresi et al. [25, 26] proposed a CMOS lab-on-a-chip for cell control and identification. The proposed lab-on-a-chip is fabricated on a standard 0.35 μm CMOS technology. This CMOS lab-on-a-chip comprises three principle parts, they are: actuation, sensing and microfluidic package. The lab-on-a-chip framework, see **Figure 4**, has two-dimensional array of microsites; these microsites provide the following functions: (1) produce the electric field which is important to make dielectrophoretic cages and (2) optically sense the existence of a single particle or cluster on the top of the microsite. Each microsite comprises of an actuation electrode, executed with a top metal plate, and underneath embedded circuitry for programming and identification. Row and column decoders can address any microsite in a random access mode, for both actuation and sensing. Particles are detected by photodiodes, embedded in the substrate that measure signal variations from uniform light impinging on the chip surface. Recently, Miled et al. presented a CMOS-based lab-on-chip system for dielectrophoresis-based cell manipulation and detection [27]. The lab-on-a-chip based on CMOS technology is fabricated utilizing a CMOS full-custom chip and a microfluidic chamber. The CMOS chip is used to pass on all parameters required to control the dielectrophoresis (DEP) parameters, for instance, electric field intensity, speed of the field and the profile of the electric field applied on the lab-on-a-chip.

2.3. Flow cytometry

Flow cytometry applications are implemented using CMOS-based lab-on-a-chip. In flow cytometry, fluorescence markers are used to tag either biological cells or single particle that is immersed in an aqueous media. These cells are hydrodynamically focused and made to traverse a small region of space across which a focused laser beam is used [28].

The input laser light will be scattered and/or fluorescence produced light will be released when a cell passes on a light source. These optical signs provided by the particles will be aggregated as spectral components of predominantly visible light. Utilizing optical multipliers and other light instrumentation, the obtained signals can be utilized to recognize and evaluate the biochemical or biophysical attributes of the cell sample.

Hartley et al. introduced a CMOS-based platform for microfluidic cytometry, where optical image detecting is incorporated on the microfluidic substrate. The stage is using flip-chip bundling innovation. **Figure 5** shows a schematic diagram of the general flip-chip digital cytometry [29].

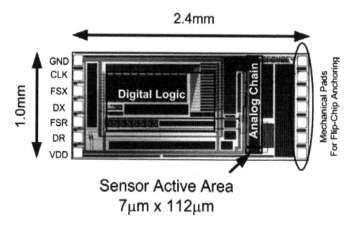

Figure 5. Photograph of the linear active pixel CMOS sensor chip with interface pads, digital logic block and analog signal path indicated [29].

As of late, Dupont et al. displayed a lab-on-a-chip that was manufactured based on silicon involving a framework of 84 diodes, i.e., single photon avalanche diodes (SPADs), to distinguish and separate fluorescently labeled single cells and fluorescent beads in a polydimethylsiloxane (PDMS) cartridge that is situated on the CMOS chip [30]. The identification depends on the diverse photon tally when either a fluorescent or a non-fluorescent bead or cell is available over a SPAD, because of the extra photons radiated from a fluorescent item. This strategy permits magnifying instrument-less fluorescence recognition and allows simple trade of the expendable microfluidic cartridge [30, 31].

2.4. Neuronal activity monitoring

Neurons are biological cells specialized in transmission and processing of information [32]. The basic neural signals are action potentials, which are transient changes of the voltage drop across the cell membrane with a typical and defined shape.

The action potential is typically 70 mV peak-to-peak. Today's standard devices utilized to measure the action potential are patch pipettes [32] or microelectrodes [33]. These apparatuses have prompted awesome change in the field of neural cell observing. Though, these strategies face downsides, for example, they require an elaborate mechanical setup, which permits observing not many cells in parallel as it were. Along these lines, they are for the most part not suitable to satisfy high-throughput prerequisites. Additionally, long-time recording is not feasible, because of the intrusive sort of contact, decreasing the life time of the cell.

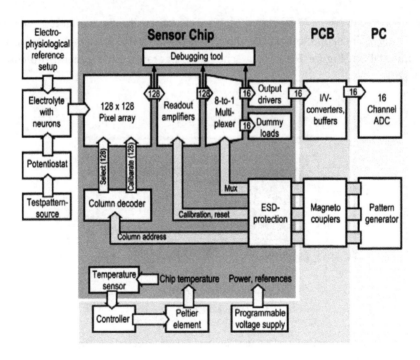

Figure 6. Architecture of sensor array with complete system [34].

Noninvasive (extracellular) recording techniques open a way to overcome these drawbacks [34–36].

In extracellular recording strategy, cells are refined straightforwardly on top of a transducing component, which is for the most part either a metallic cathode or a floating gate transistor [34, 35]. At the point when an action potential happens in a cell, the nearby stream of particles all through the cell causes the membrane to be polarized in a non-uniform way. The ionic current streams over the cleft resistance and sets up an electric field which prompts electrical charge in the fundamental transducer, which is the recorded signal.

Eversmann et al. proposed a CMOS lab-on-a-chip for non-invasive recording and detection of neuronal activity [34, 35]. **Figure 6** shows the system setup and chip architecture. Heer et al. [37] presented a CMOS lab-on-a-chip for bidirectional communication (stimulation and recording). **Figure 7** shows a picture of the CMOS lab-on-a-chip. **Figure 8** shows a schematic of the overall system.

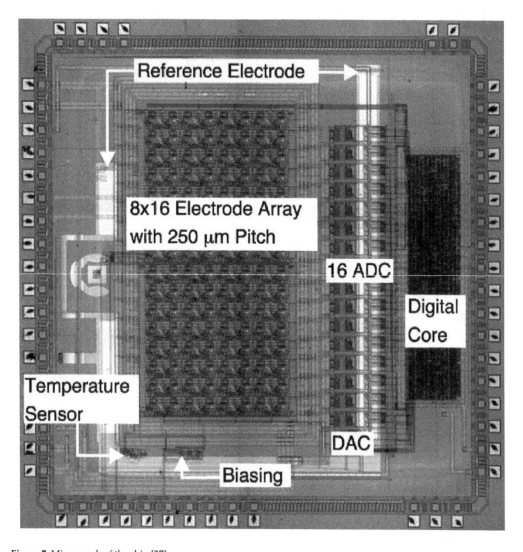

Figure 7. Micrograph of the chip [37].

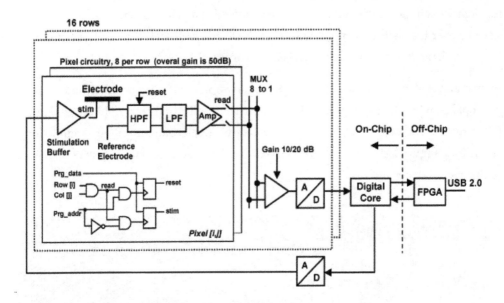

Figure 8. Block schematic of the overall system architecture [37].

Recently, Rothberg et al. [38] presented a CMOS sensor chip which can be used to harbor neurons and glial cells in vitro. The CMOS chip incorporates a vast cluster of sensor components, each with a single floating gate connected to an underlying ion-sensitive field-effect transistor (ISFET), to gauge the pH [39].

2.5. Electric field imager

Ghallab et al. [12, 40] presented a CMOS-based lab-on-a-chip. The used technology is CMOS 180 nm TSMC technology. This lab-on-a-chip is utilized to get electric field images for biocells. These images depend on the electrical characteristics of the cells which can be utilized to distinguish between various sorts of cells [12]. **Figure 9** shows the die picture; the total die area is 0.75 × 0.6 mm. The CMOS-based lab-on-a-chip includes two main parts. They are: (1) the actuation part, that generates the planned non-uniform electric field pattern; the actuation part has four electrodes that generate a DEP force to levitate the cell that should be identified; (2) the sensing part, which is an array of the differential electric field sensitive field effect transistor (DeFET) sensor [12, 40, 41].

2.5.1. The actuation part

The actuation electrodes are the quadrupole setup, see **Figure 9**. Utilizing this arrangement, the profile of the non-uniform electric field can be controlled by associating the entire four electrodes or few of them. Likewise, the quadrupole levitator contains an azimuthally symmetric electrode arrangement fit for supporting passive stable molecule levitation [42].

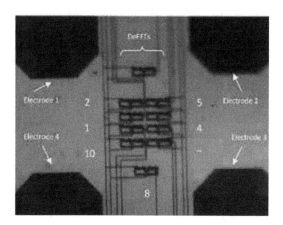

Figure 9. The Die picture shows the quadrupole electrodes and the DeFET sensors [42].

2.5.2. The sensing part

The sensing part is made out of an array of DeFETs. This array is situated around the center of the chip, where the cell is levitated (suspended). In DEP levitation, the controlling electric field is a non-uniform electric field. Consequently, we can recognize the electric field by utilizing the electric field sensitive field effect transistor [eFET] as an electric field sensor.

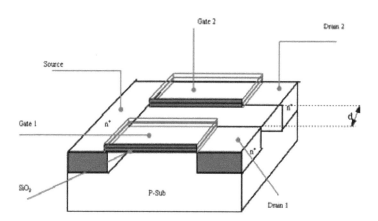

Figure 10. Physical structure of an eFET [12, 42].

Figure 10 demonstrates the physical structure of the eFET. It comprises of two neighboring drains, two nearby floating gates with separation space "d" between the two floating gates, and one source.

For the eFET, it is proportionate to two indistinguishable enhancement MOSFET transistors. Affected by the non-uniform electric field, different current passes through each drain. Because of the drain current reliance on the gate voltage, the eFET can sense the difference between the two gate voltages, which mirrors the magnitude of the used non-uniform electric field. To increase the measurement range of the eFET, CMOS technology is used to implement the differential electric field sensitive field effect transistor (DeFET) sensor, and this sensor will be the basic sensor in the sensing part of the proposed electric field imager [40–42].

2.5.3. DeFET sensor

The DeFET consists of two complementary eFETs, one is a P eFET and the second one is an N eFET [12, 42]. The equivalent circuit of the DeFET is shown in **Figure 11**.

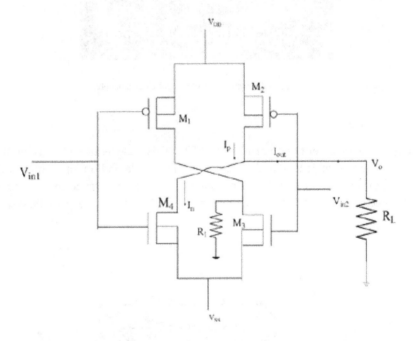

Figure 11. An equivalent circuit of a DeFET [12], the two gates of P eFET and N eFET are connected with each other, and there is a cross coupling between the two drains of the P eFET and the N eFET. The output current I_{Out} is equal to the difference between the two drain currents $I_p - I_n$ (i.e., $I_{Out} = I_p - I_n$). I_p and I_n are functions of the two applied gate voltages V_{in1} and V_{in2}, respectively.

The DeFET is intended to accomplish a voltage V_{Out}, specifically identified with the difference between the two gate voltages ($V_{in1} - V_{in2}$), and $V_{in1} - V_{in2}$ is equivalent to the generated electric field over the two gates (E) (E) multiplied by the distance between them ($V_{in1} - V_{in2}/Y = E$), where Y is the distance between the two split gates, which is constant. Thus, V_{Out} is related straightforwardly to the non-uniform electric field's intensity. In this way by measuring V_{Out}, the non-uniform electric field intensity can be obtained.

To extract an expression that relates V_{Out} and E, and from **Figure 11**, the output current (I_{out}) is:

$$I_{out} = I_p - I_n$$

(1)

The sensitivity is given by

$$S = \frac{dI_{out}}{dE} = g_m Y \tag{2}$$

where g_m is the transconductances of the transistors M2 and M4, which have equal g_m and Y is the distance between the two gates.

From (2)

$$I_{out} = SE + \text{Constant} \tag{3}$$

The output voltage (V_{out}) can be expressed in terms of the sensitivity and the electric field:

$$V_{out} = I_{out} R_L = SE R_L + V_{Uni} \tag{4}$$

where V_{uni} is the output voltage due to a uniform electric field (i.e., $V_{in1} = V_{in2}$), and R_L is the load resistance. From (2) into (4)

$$V_o = g_m Y R_L E + V_{Uni} \tag{5}$$

From equation (5), it can be observed that V_o is related linearly to the intensity of the non-uniform electric field generated on the top of the DeFET sensor. Thus, in an array form of DeFET sensors, an image related to the electric field intensity at different locations can be extracted.

2.5.4. Experimental results

The CMOS electric field imager is experimentally examined with micro-beads having a relative permittivity of 2.5 and radius of 4.5 and 10.45 μm, respectively. The experiment was done in two steps. First, the chip is tested on air. Second, it is tested with micro-beads suspended on medium. This CMOS electric field imager allows sensing the impact of the cells on the electric field intensity profile.

Figures 12 and **13** show sample of the results. The configuration used is as follows: (1) a 3V sinusoidal signal with 3 MHz frequency is connected to Electrode 2 and Electrode 4, see **Figure 9**; (2) an out of phase (−3V) sinusoidal signal with 3 MHz is connected to Electrode 1 and Electrode 3. The cells were suspended in de-ionized water with a measured conductivity between 1.3 and 1.9 μs/cm. A microscope with digital camera is used to observe the particles from the top.

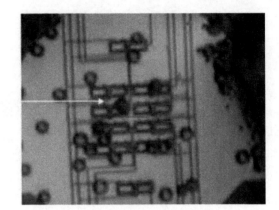

Levitated cell

Figure 12. Levitated polystyrene cell with diameter 8.9 μm [42].

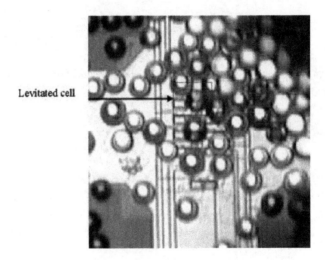

Levitated cell

Figure 13. Levitated polystyrene cell with diameter 20.9 μm [42].

Figures 12 and **13** demonstrate a trapped cell not long after the electric field is turned on, simply over the center of the imager. The levitated/trapped cell shows up out of focus of the microscope's focal length [42]. **Figure 14** shows the measured output voltage from the DeFET sensor numbers 1, 4, 7 and 10, in air and after adding the suspending fluid, which contains 4.5 and 10.5 μm radius polystyrene microspheres. From this figure, it can be noticed that DeFETs 4 and 7 show lower readings compared with DeFETs 1 and 10. Also, note from **Figure 14** that for the cells with 20.9 μm diameters, DeFETs 1, 4 and 10 provide high readings, while DeFET 7 does not. Leading to the surmise that there are cells above 1, 4, and 10, but no cells above DeFET 7 [42].

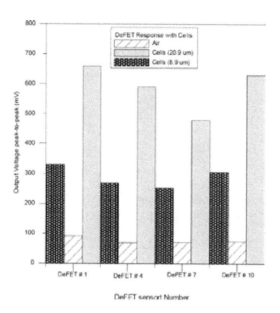

Figure 14. The DeFET sensors' response in air and in fluid contains different cell size [42].

3. Conclusion

In this chapter, up-to-date, advances in CMOS circuits and systems based lab-on-a-chip are provided and discussed. Additionally, distinctive applications of CMOS lab-on-a-chip are presented. CMOS-based lab-on-a-chip guarantees many advantages to be provided to the biology and medicine fields. These advantages are portability, disposability and miniaturization. Also, these advantages allow scientists to do complex experiments anywhere with portable devices. However, until now, there is an unmet need for lab-on-a-chip to effectively deal with the biological systems at the cell level.

Acknowledgements

This research was partially funded by Zewail City of Science and Technology, AUC, the STDF, Intel, SRC, Mentor Graphics, ASRT and MCIT.

Author details

Yehya H. Ghallab[1,2*] and Yehea Ismail[2*]

*Address all correspondence to: yghallab@zewailcity.edu.eg and yismail@zewailcity.edu.eg

1 Department of Biomedical Engineering, Helwan University, Cairo, Egypt

2 Centre of Nanoelectronics and Devices (CND) at Zewail City of Science and Technology/
American University in Cairo, Cairo, Egypt

References

[1] Huang Y. et al. Lab-on-CMOS integration of microfluidics and electrochemical sensors. Lab Chip. 2013; 13: 3929–3934. DOI: 10.1039/C3LC50437A

[2] Huang C, Huang Y, Yen P, Tsai H, Liao H, Juang Y, Lu S, Lin C. A CMOS wireless biomolecular sensing system-on-chip based on polysilicon nanowire technology. Lab Chip. 2013; 13: 4451–4459. DOI: 10.1039/C3LC50798J

[3] Jackson S, Cole D. Graduate global public health education: activities and outcomes in relation to student prior experience. Global Journal of Health Science; 2013; 5:54–63 DOI: 10.3402/gha.v7.24526

[4] Yager P, Edwards T, Fu E, Helton K, Nelson K, Tam M, Weigl B. Microfluidic diagnostic technologies for global public health. Nature. 2006; 442: 412–418. DOI:10.1038/nature05064

[5] UNAIDS Report 2014, http://www.unaids.org/en available on line [internet]

[6] Pratt G. et al. Monitoring HIV drug adherence with a paper-based assay. IEEE Point-of-Care Healthcare Technologies (PHT): IEEE; 2013; 69–71

[7] Craig I, Xiaohua X. Can HIV/AIDS be controlled? Applying control engineering concepts outside traditional fields. IEEE Magazine for Control Systems. 2005; 25: 80–83. DOI: 10.1109/MCS.2005.1388805

[8] Welch D, Blain Christen J. Seamless integration of CMOS and microfluidics using flip chip bonding. Journal of Micromechanics and Microengineering, 2013; 23.035009

[9] Mudanyal O. et al. Wide-field optical detection of nanoparticles using on-chip micro-scopy and self-assembled nanolenses. Nature Photonics. 2013; 7: 247–254. DOI: 10.1038

[10] Adam T. et al. Nano lab-on-chip systems for biomedical and environmental monitoring. African Journal of Biotechnology. 2013; 12: 5486–5495. DOI: 10.5897/AJB11.3826

[11] Ghallab Y H, Ismail Y. CMOS based lab-on-a-chip: applications, challenges and future trends. IEEE Circuit and Systems Magazine.2014; Q2: 27–47.

[12] Ghallab Y H, Badawy W. Lab-on-a-Chip: Techniques, Circuits and Biomedical Applications. Artech House Publisher, USA, 2010. ISBN: 978-1-59693-418-4.

[13] Gulari M N, Ghannad-rezaie M, and Chronis N. "A compact, optofluidic system for measuring red blood cell concentration", IEEE conference on Solid- State Sensors, Actuators and Microsystems. 2013; 2552–2555.

[14] Amine M. et al. Dielectrophoresis-based integrated lab-on-chip for nano and micro-particles manipulation and capacitive detection. IEEE Transactions on Biomedical Circuits and Systems. 2012; 6: 120–132. DOI: 10.1109/TBCAS.2012.2185844

[15] Esfandyarpour R. et al. Simulation and fabrication of a new novel 3D injectable biosensor for high throughput genomics and proteomics in a lab-on-a-chip device. Nanotechnology; 2013; 24, DOI: 10.1088/0957-4484/24/46/465301

[16] Adam T. et al. Nano lab-on-chip systems for biomedical and environmental monitoring. African Journal of Biotechnology. 2013; 12: 5486–5495. DOI: 10.5897/AJB11.3826

[17] Huang Y. et al. Lab-on-CMOS integration of microfluidics and electrochemical sensors. Lab Chip. 2013; 13: 3929–3934. DOI: 10.1039/C3LC50437A

[18] Alrifaiy A. et al. Polymer-based microfluidic devices for pharmacy, biology and tissue engineering. Polymers; 2012: 4, 1349–1398. DOI: 10.3390/polym4031349

[19] Park D. et al. Titer plate formatted continuous flow thermal reactors for high throughput applications: fabrication and testing. Journal of Micromechanics and Microengineering; 2010: 20,291–300 DOI:10.1088/0960

[20] Myong J. et al. Detection of bacterial pathogen DNA using an Integrated complementary metal oxide semiconductor microchip system with capillary array electrophoresis. Journal of Chromatography B. 2003; 783: 501–508. DOI: 1570-0232/02

[21] Cheng S. et al. Effective amplification of long targets from cloned inserts and human genomic DNA. PNAS. 1994; 91: 5695–5699.

[22] Norian H. et al. Integrated CMOS quantitative polymerase chain reaction lab-on-chip. Symposium on VLSI Circuits (VLSIC).2013. IEEE>. C220–C221.

[23] Lee J. et al. Electrowetting and Dielectrowetting-on-dielectric for microscale liquid handling. Sensors and Actuators A. 2002; 95: 259–268. DOI: 10.1016/S0924-4247(01)00734-8

[24] Sawan M. et al. Cmos/microfluidic lab on- chip for cells-based diagnostic tools. Engineering in Medicine and Biology Society (EMBC), 2010 Annual International Conference of the IEEE, 2010. IEEE. 5334–5337.

[25] Medoro G. et al. Lab on a chip for live-cell manipulation. IEEE Design & Test of Computers. 2007; 24: 26–36. DOI: 10.1109/MDT.2007.21

[26] Manaresi N. et al. A CMOS chip for individual cell manipulation and detection. IEEE Journal of Solid State Circuits. 2003; 38: 2297–2305. DOI: 10.1109/JSSC.2003.819171

[27] Amine M. et al. Dielectrophoresis-based integrated lab-on-chip for nano and micro-particles manipulation and capacitive detection. IEEE Transactions on Biomedical Circuits and Systems. 2012; 6: 120–132. DOI: 10.1109/TBCAS.2012.2185844

[28] Wang Z. Measurements of scattered light on a microchip flow cytometer with inte-grated polymer based optical elements. Lab. Chip.2004; 4: 372–377. DOI: 0.1039/b400663a

[29] Hartley L. et al. Hybrid integration of an active pixel sensor and microfluidics for cytometry on a chip. IEEE Transactions on Circuits and Systems — I. 2007; 54: 99–110. DOI: 10.1109/TCSI.2006.887456

[30] Duponta E. P. et al. Fluorescent magnetic bead and cell differentiation/counting using a CMOS SPAD matrix. Sensors and Actuators B. 2012; 174: 609–615. DOI: 10.1016/j.snb.2012.06.049

[31] Gulari M. N. et al. A compact, optofluidic system for measuring red blood cell concen-tration. 2013. IEEE Conference on Solid-State Sensors, Actuators and Microsystems, IEEE. 2552–2555.

[32] Nicholls J. G. et al. From Neuron to Brain, 3rd ed. Sinauer Assoc. Inc., Sunderland, 1992.

[33] Purves R. D. Microelectrode Methods for Intracellular Recording and Ionophoresis. Academic, London, U.K., 1981.

[34] Eversmann B. et al. A 128 x 128 CMOS biosensor array for extracellular recording of neural activity. IEEE Journal of Solid State Circuits. 2003; 38: 2306–2317. DOI: 10.1109/JSSC.2003.819174

[35] Eversmann B. et al. CMOS sensor array for electrical imaging of neuronal activity. IEEE International Symposium on Circuits and Systems (ISCAS 2005), Kobe, Japan. IEEE; 2005, 3479–3482.

[36] Rothberg J. M. et al. An integrated semiconductor device enabling non-optical genome sequencing. Nature; 2011: 475 (7356): 348–352. DOI: 10.1038/nature10242

[37] Heer F. et. al. CMOS microelectrode array for bidirectional interaction with neuronal networks. IEEE Journal of Solid State Circuits; 2006: 41: 1620–1629. DOI: 10.1109/JSSC.2006.873677

[38] Rothberg J. M. et al. An integrated semiconductor device enabling non-optical genome sequencing. Nature; 2011:475 (7356): 348–352. DOI: 10.1038/nature10242

[39] Soea A. K. et al. Neuroscience goes on a chip. Biosensors and Bioelectronics. 2012; 35: 1–13.

[40] Ghallab Y. H., Badawy W. Magnetic Field Imaging Detection Apparatus, US Patent, 2008: 7, 405, 562.

[41] Ghallab Y. H., Abdelhamid H., Ismail Y. Lab-on-a-chip based on CMOS technology: system architectures, microfluidic packaging and challenges. IEEE Test and Design; 2015: 32: 20–31.

[42] Ghallab Y. H., Badawy W. A single CMOS chip for biocell tracking, levitation, detection, and characterization. International Symposium on Circuit and System (ISCAS 06), Greece. IEEE; 2006, 3349–3352.

8

Digital Nucleic Acid Detection Based on Microfluidic Lab-on-a-Chip Devices

Xiong Ding and Ying Mu

Abstract

Microfluidic lab-on-a-chip (LOC) technologies have been developed as a promising alternative to traditional central laboratory-based analysis approaches over several decades due to the capability of realizing miniaturized multiphase and multistep reactions. In the field of nucleic acid (NA) diagnosis, digital NA detection (dNAD) as a single-molecular-level detection is greatly attributed to the perfect combination of NA amplification and microfluidic LOC techniques. In this chapter, the principle, classification, advances, and application of dNAD will be involved. In particular, the focus will be on chip-based dNAD for giving a deep interpretation of the analysis and evaluation of digital detection. The future prospect of dNAD is also anticipated. It is sure that dNAD by means of microfluidic LOC devices as the promising technique will better serve the ambitious plan of precision medicine through absolute quantitation of NA from individuals.

Keywords: digital nucleic acid detection, lab-on-a-chip, microfluidic chip, digital PCR, quantitation

1. Introduction

It has been clearly investigated that nearly all of the diseases possess a series of biomarkers associated with nucleic acid (NA) molecules during the development of biological researches [1–5]. Determining these NA molecules and their intercellular and extracellular changes is a well-worked strategy for estimating therapy efficacy, monitoring minimal residual diseases, unveiling the mechanisms of cellular signal transduction, and so on [6–8]. To reflect individual genetic differences, single-molecule level quantitation of NAs has been increasingly concerned recently due to its superiority on analytical sensitivity and accuracy [9, 10]. Furthermore, single NA molecule detection is also highly preferred as the calibration strategy for next-generation sequencing (NGS) to better serve recently proposed ambitious plans of precision medicine [11–13]. Thus, testing NAs, especially in single-molecule level, plays an essential role in modern biological researches and diagnosis fields.

At present, the widely used approaches in detecting NA molecules are quantitative polymerase chain reaction (qPCR) and quantitative reverse transcription-PCR (qRT-PCR). Apart from the capability of real-time monitoring the amplification, they can be applied to quantify the target NA molecules through two common strategies: relative quantification and absolute quantification. The former is based on internal reference genes (namely, housekeeping genes) to normalize and reflect fold differences in expression levels of mRNA, which is commonly interpreted as cDNA [14, 15]. The latter can provide the exact number of targeted molecules using an established standard curve of the change in quantification cycles with known molecule number of NA standards [16–19]. However, qPCR is compromising the ability of single-molecule quantitation analysis [20, 21]. Alternatively, when PCR meets microfluidic or nanofluidic chips, a highly sensitive NA quantification technique [digital PCR (dPCR)] emerges, estimating NAs advantageously at a single-molecule analysis level [22].

At the end of 20th century, the first concept of dPCR was proposed by Vogelstein and Kinzler [23]. Since the concept was proposed, many dPCR platforms have been launched for several

Figure 1. Some vendors and their launched microfluidic chips and dPCR devices.

decades based on differently designed microsystems, including femtoliter array, spinning disk, SlipChip, droplet, microfluidic formats, and so on [24–30]. Some even have been successfully paced into industrial phase because of the superiority of testing and the promising application. Currently, several vendors in the biological industry, such as Fluidigm, Bio-Rad Laboratories, Life Technologies (ThermoFisher), RainDance Technologies, and Formulatrix, have launch individually their commercialized dPCR devices (**Figure 1**).

Apart from dPCR, digital isothermal NA amplification (dINAA) devices also arouse great concern. Unlike dPCR, dINAA leans on the isothermal NA amplification, which can be carried out at a consistent temperature, obviating the requirement of highly stable thermocycling devices. Thus, when targeting practical point-of-care testing (POCT) devices, dINAA is superior to dPCR. However, viewed from the principle of realizing digital detection, the concept of dINAA is the same as that of dPCR, just replacing PCR with isothermal amplification. In particular, due to loop-mediated isothermal amplification (LAMP) displaying as the best promising method among a lot of isothermal NA amplifications, digital LAMP (dLAMP) is the first dINAA developed [31, 32]. Later, other dINAAs have been reported, such as digital multiple displacement amplification (dMDA), digital isothermal multiple-self-matching-initiated amplification (dIMSA), digital recombinant polymerase amplification (dRPA), and so on [33–38]. However, the development of dINAA devices is still in the research stage, as the commercial products have not been launched yet.

As of now, more and more researchers are enthusiastic about the potential of digital NA detection (dNAD) based on microfluidic Lab-on-a-Chip (LOC) devices, since an increasingly significant role has been played in single-cell analysis, early diagnosis of cancer, prenatal diagnosis, and so on. In this chapter, we will concentrate on the principle, classification and advances, analysis and evaluation, application, and future prospects of dNADs that are accomplished either through commercialized LOC devices or the devices our laboratory or other laboratories have established.

2. Principle of dNAD

According to the strategies of amplifying NA, dNAD or single-molecule NA detection can be divided into dPCR and dINAA. However, both of them share the same principle.

Generally speaking, the principle of dNAD is composed of three core steps [39]. First, the original sample should be partitioned into thousands or hundreds of thousands of individual microreactions, endeavoring to make each contain nearly one target molecule. Second, the number of "positive" microreactors indicated either in a real-time reaction or in an endpoint reaction is counted and analyzed. Third, the concentration of nonpartitioned sample is calculated using certain statistical methods. In theory, if the number of microreaction is more enough or the number of target molecules is less enough, one reaction unit with positive signal represents one target molecule. However, in fact, a positive partition may contain more than one molecule. Therefore, in calculating the target's true concentration, a Poisson distribution is adopted in hope to correct the results. Therefore, dNAD can be considered as a binary output (present or absent like "1" or "0" in computer science) measurement, giving a direct and high-confidence NA molecule's measurement method [10, 40]. Compared to conventional tube-based NA detection, digital analysis is superior in realizing the absolute quantification with high sensitivity, high precision, and low ambiguity, avoiding the requirement of establishing a standard curve.

3. Classification and advances of dNAD

In the early stage of digital detection, the used materials are 96- and 384-microwell plates. Then, due to the rapid development of microfluidic chip techniques, an increasing number of digital detection devices emerge. Also, a variety of materials have been used individually or jointly, such as silicon wafer, quartz, glass, polydimethylsiloxane (PDMS), polymethyl methacrylate, and so on. According to the approaches to partition reaction mixture, the currently launched dNAD methods can be roughly grouped into three categories: plate-based dNAD (pdNAD), droplet-based dNAD (ddNAD), and chip-based dNAD (cdNAD). On structural design, each classified dNAD has the advantages and disadvantages, and the corresponding commercial devices are also developed. In this subchapter, we are going to narrate their features and recent advances in either commercial or research aspects.

3.1. pdNAD

At present, most of the pdNADs are established as dPCR devices, but they are not hard to be developed as dINAA platforms. As the first generation of dNAD, plate-based dPCR (pdPCR) was first conducted using plenty of commercially available 96- and 384-microwell plates [23, 41]. The biggest benefit for this kind of digital platform is saving to create the plates that have been widely used in conventional PCR. Each microwell undertakes each microreaction; therefore, the high sensitivity and accuracy of detection lean entirely on the enough number of microwells. However, actually, the number is hard to be reached just using microwell plates.

Another problem causing the embarrassment is the volume of reagents required [42]. For each microwell, more than 5 μL are needed, and the cost of reagents inevitably daunts most researchers, let alone the application for POCT. To break the barriers, some researchers made modification. As shown in **Figure 2**, Morrison et al. deceased the volume of microreaction into 33 nL using a stainless steel plate (25 mm in width and 75 mm in length) in which up to 3072 microholes (320 μm in diameter) were created [43]. In contrast, the required volume was reduced to 1/64, and the throughput was increased by 24-fold, although it had the comparable sensitivity to the past. At present, this technique has been applied to commercial devices in 2009, the OpenArray RealTime PCR System from Life Technologies. However, as the number of reaction units increases, the problem turns into how to efficiently load the reagents. Consequently, it has to use some ancillary equipment-like microarray spotter or mechanical arms, which in turn raises the cost and is cumbersome.

Considering the embarrassing situation, in the second half of 2013, Life Technologies launched the next-generation digital detection device, the QuantStudio 3D dPCR system [42, 44]. It is a simple and affordable platform to provide the reliable and robust dPCR. The device used a

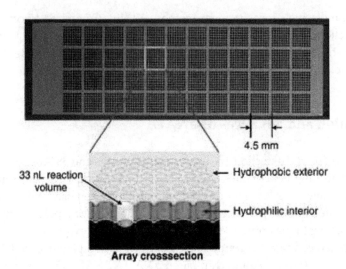

Figure 2. Plate-based chip used for the OpenArray RealTime PCR System. A rectilinear array of 3072 microholes with 320 μm in diameter was fabricated in a stainless steel plate (25×75×0.3 mm). The volume of each hole was approximately 33 nanoliers, and to match the pitch of the wells in a 384-well microplate, the 48 groups of 64 holes are spaced at 4.5 mm.

special plate (10 mm in width and 10 mm in length) where a total of 20,000 hexagonal microwells are fabricated. The volume of each microwell is 0.8 nL, and each reaction well is isolated absolutely from its neighbors. At present, the system has been applied to the absolute quantification of viral load, low-level pathogen detection, sensitive genetically modified organism (GMO) detection, differential gene expression, copy number variation (CNV), NGS library quantification, and rare mutation analysis [45–50]. Although the cost of reagents is reduced, the system still calls for supporting instruments to load the reagents, amplify the sample, and read the results.

For high-throughput sample analysis, 96- and 384-microplate formats are still of use. Formulatrix introduced a new commercial high-throughput pdPCR device termed as constellation dPCR. The device brings the digital analysis to a 96-sample microplate format, and the so-called high-throughput results from the preformation of dPCR on 96 samples at once and up to 384 samples per hour. As required, the number of partitions for each microwell in the plate can be easily increased, and it reaches 496 for the 96-microplate format.

3.2. ddNAD

ddNAD can go back to emulsion PCR (ePCR) [51–54]. ePCR is widely used for NGS (**Figure 3**) [55]. After generating a DNA library, the fragments of genomic DNA are attached to the beads, because their surface is modified with oligonucleotide probes whose sequences are complementary to the sequences of the fragments. When the beads are compartmentalized into water (the PCR reagent)-oil emulsion droplets, plenty of microreactors are produced. Since each bead captures single-stranded DNA fragment, in theory, ePCR can amplify it down to one DNA molecule. However, it is not easy to partition the fragments and beads into one droplet simultaneously, and then the performance of ePCR suffers from variety. Benefitting from the rapid development of microfluidic LOC techniques, ddNAD also enjoys a huge boom in recent years. dPCR is still the main part in ddNAD, but droplet-based dINAAs including dLAMP, dRPA, digital rolling circle amplification (RCA), and digital hyperbranched RCA (HRCA) are showing up more and more [34–37, 56].

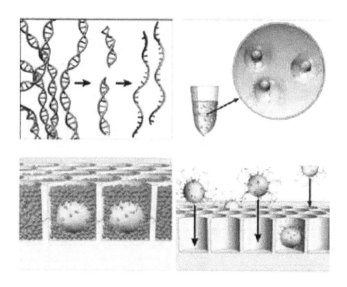

Figure 3. ePCR used for NGS. Top left: The genomic DNA is isolated, fragmented, ligated to adapters, and separated into single strands. Top right: Fragments are bound to beads that are captured in the droplets of a PCR mixture-in-oil emulsion. Then, ePCR occurs within each droplet. Bottom right: After breaking emulsion and denaturing the DNA strands, beads with single-stranded DNA are deposited into wells of a fiber-optic slide. Bottom left: Pyrophosphate sequencing is initiated within each well after depositing smaller beads carrying immobilized required enzymes.

Beer et al. successfully created picoliter-scale water-in-oil droplets by using a shearing T-junction in a fused-silica device in 2008 [57]. The NA used for the device was RNA; therefore, an off-chip valving system was integrated to stop the droplet motion, because a different thermal cycling was required for reverse transcription and subsequent PCR amplification. Each droplet contained the PCR mixture of single-copy template, primers, and reaction buffer, which was really termed as digital detection. One year later, Mazutis et al. developed a method for high-throughput dINAA platform in a 2 pL droplet-based microfluidic system [35]. The isothermal HRCA was used to perform the DNA amplification in droplets. This platform was demonstrated to allow fast and accurate digital quantification of the template. In 2011, Zhong et al. reported another picoliter-scale droplet-based multiplexing dPCR platform, breaking the one target per color barrier of qPCR [58]. The number of droplets generated reached more than 10^6, which was enough for enhancing the likelihood that only one DNA molecule was amplified in each droplet. Given its great potential in application, RainDance Technologies launched the commercial digital detection system with the highest droplet throughout, the RainDrop dPCR system. Unfortunately, the system may consume up to 50 µL reagents per sample. Considering this point, the system may be not proper for rare sample detection. At the same year, Hindson et al. also established a high-throughput droplet-based dPCR (ddPCR) platform [59]. A total of 2 million droplets were generated, and the droplets were then transferred into a 96-well plate for TaqMan probe-based PCR. Finally, to read out the results, a flow cytometry-like double-channel fluorescence detection device was used in a microfluidic chip, in which droplets went through one by one. The platform was confirmed to realize the accurate measurement of germ-line CNV, discriminate the mutant molecules from the wild molecules with 10^5-fold excess, and absolutely quantify circulating fetal and maternal DNA from cell-free plasma. Based on the platform, the first commercial ddPCR system was launched by QuantaLife in 2011, but in the end of that year Bio-Rad Laboratories purchased the company and launched the QX100 ddPCR. Recently, the new version, QX200 ddPCR system, is also available.

Apart from the process of droplet generation and subsequent NA amplification, other approaches to generate droplet are also reported. As shown in **Figure 4**, Shen et al. described a SlipChip to create droplet array [26]. The SlipChip was composed of two glass plates, in which elongated wells were designed to overlap and form the fluidic path for reagent loading. After sample loading, the simple slipping of the two plates broke the path, removing the overlap among wells and generating 1280 droplet array (2.6 nL for each). The device had a reservoir preloaded with oil, so each microreactor was absolutely isolated from each other

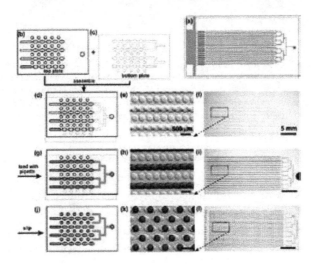

Figure 4. Design and mechanism of the SlipChip for dPCR. The top plate is outlined with a black solid line, the bottom plate is outlined with a blue dotted line, and red represents the sample. (a) Schematic drawing shows the design of the entire assembled SlipChip for dPCR after slipping. (b) Schematic drawing of part of the top plate. (c) Schematic drawing of part of the bottom plate. (d–f) The SlipChip was assembled such that the elongated wells in the top and bottom plates overlapped to form a continuous fluidic path. (g–i) The aqueous reagent (red) was injected into SlipChip and filled the chip through the connected elongated wells. (j–l) The bottom plate was slipped relative to the top plate such that the fluidic path was broken up and the circular wells were overlaid with the elongated wells, and aqueous droplets were formed in each compartment. (d, g, and j) Schematic of the SlipChip. (e, h, and k) Zoomed-in microphotograph of the SlipChip. (f, i, and l) Microphotograph of the entire SlipChip.

during thermal cycling. Finally, the results were read out using endpoint fluorescence intensity. The biggest advantage of SlipChip is the capability of realizing multistep manipulation of plenty of microvolumes to form droplet array in parallel. Attributed to the remarkable feature, until now, the SlipChip has been applied to perform immunoassays, protein crystallization, multiplex PCR, dPCR, dLAMP, dRPA, and so on [26, 34, 60–64].

To make the droplet more stable and to easily collect the amplified products, Leng et al. invented an agarose droplet-based single-molecule ePCR device [51]. The agarose performed the unique thermoresponsive sol-gel switching property, and a microfluidic chip was designed to produce uniform agarose solution droplets. Schuler et al. applied centrifugal step emulsification to the fast and easy generation of monodisperse droplets [37]. Only by adjusting the nozzle geometry (depth, width, and step size) and interfacial tensions droplets with desirable diameters could be produced. Using this droplet device, dRPA was successfully established for the absolute quantification of *Listeria monocytogenes* DNA concentration standards within 30 min.

In ddNAD, the microreactors are generated by carefully titrating emulsions of water, oil, and chemical stabilizer; therefore, there is no requirement of the walls of microwells to separate the microreactors. Compared to pdNAD, ddNAD can easily achieve higher throughput via a microdroplet generator to produce hundreds of thousands of droplet reactions per sample. However, the workflow of ddPCR is complicated, referring to generating droplet, transferring droplet, sealing microplate, conventional PCR, and reading out the signal by other devices.

3.3. cdNAD

The development of cdNAD is greatly attributed to the rapid progress of microfluidic techniques, which can realize the low cost, low volume, and high-throughout paralleled NA detections. In the last several decades, microreactors in cdNAD are mainly formed either by the mechanical compartmentalization of PDMS or by the succeeding isolation via immiscible liquid phase. In particular, for PDMS-based chips, the establishment of multilayer soft lithography (MSL) techniques developed by Unger et al. in 2000 also gives a huge boost, making the high-density microwells, micropumps, and microvalves easily fabricated [65]. Based on different power sources to partition reagents, cdNAD can be divided into three categories: integrated fluidic circuit (IFC) cdNAD, self-priming compartmentalization (SPC) cdNAD, and localized temporary negative pressure (LTNP)-assisted cdNAD as well as other cdNADs.

3.3.1. IFC cdNAD

The outstanding feature of IFC chip is the special design of separated and interlaced liquid and gas channels, as shown in **Figure 5**. Taking advantage of the high elasticity of PDMS, hundreds or thousands of microreaction units are formed rapidly when gas channels are added with pressure.

In 2006, Ottesen et al. used the IFC chip to achieve dPCR analysis [66]. A total of 1176 microreaction units were produced by controlling accurately the integrated microvalves and

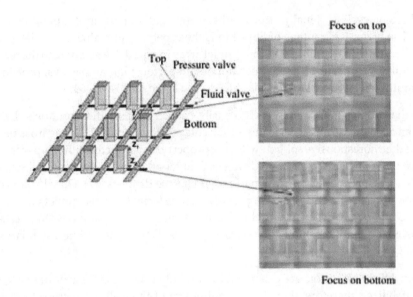

Figure 5. An IFC chip-based 12×765 digital array from the Fluidigm. Left: Schematic diagram of a part of the IFC chip in which microchambers were connected and isolated by fluidic channels and pressure lines. Right: Optical microscopic images of the part.

removing the conventional microarray spotter and plates. Now, the IFC chip-based dPCR (IFC cdPCR) platform is successfully established for commercial purpose by Fluidigm. As the first vendor to commercialize dPCR device, Fluidigm provides two IFC-based systems, the BioMark HD and EP1 systems. In the two systems, the PCR reagents are mixed and partitioned automatically, the thermocycling is integrated, and the results can be read out after reaction. The BioMark HD system can offer real-time detection for each tiny reaction and eliminate false positives according to the data, so the system is also available to qPCR. Compared to BioMark HD, EP1 is just an endpoint detection machine, giving the binary output-like data whether or not the microreaction occurs. Recently, Fluidigm also combines with Olink to detect human protein biomarkers based on proximity extension assay (PEA) technology. Until now, the dPCR device from Fluidigm has been applied to single-cell analysis, early diagnosis of cancer, and prenatal diagnosis.

In 2011, Heyries et al. developed a megapixel dPCR platform in which 10^6 microunits were fabricated, and the microreaction's volume reached down to 10 pL (**Figure 6**) [67]. The density of the microreactors reached up to 440,000/mm^2, which was the highest density for IFC platform. On detection performance, this device was able to discriminate one mutant molecule from 10^5 wild molecules and achieve the discrimination of a 1% difference in chromosome copy number. After the platform, in 2012, Men et al. published anther dPCR platform possessing the lowest volume (36 fL) of microreactors until now. Its density of microreactors was more than 20,000/mm^2 [24]. After loading the reagents into all microreactors simultaneously, the deformation of a PDMS membrane was used to completely seal the filled microreactors. Due to the femtoliter-level microreactors fabricated, the device can greatly reduce the consumption of reagent and sample.

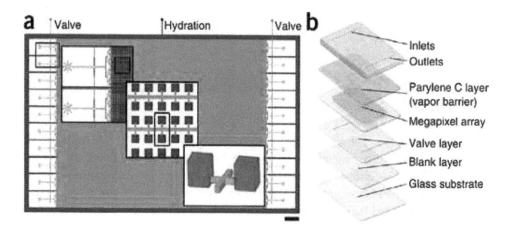

Figure 6. Schematic of megapixel dPCR device (a) and the layered device structure (b).

For dINAA, IFC chip is also combined with isothermal MDA to develop dMDA for enumeration of total NA contamination [33]. On detection of microbial genomic DNA fragments, dMDA performs higher sensitivity with orders of magnitude than qPCR.

By making the microchamber smaller or increasing its number, IFC cdNAD has a potential to be developed into a digital detection platform with higher throughout, higher density, and higher discrimination ability, but this platform still relies on the control system of integrated microvalves and micropumps to load and partition the reagents, which is hard to be applied towards POCT. Furthermore, narrowing the size of microchamber endlessly may have an impact on the efficiency of NA amplification.

3.3.2. SPC cdNAD

Targeting practical POCT devices, currently proposed plate-based, droplet-based, and IFC cdNADs are confronted with the huge challenge of demand for peripheral control instrument, for instance, external syringe pumps, droplet generation devices, and plenty of integrated microvalves and micropumps. Upon this challenge, the built-in power-driving, self-partitioning, easy-to-use, and low-priced SPC cdNADs were developed by our laboratory. The built-in power results from the gas solubility and permeability of PDMS, because PDMS remains absorbing gas and letting gas go through them, although PDMS is in a solid state in chips [68].

The chip possesses the prominent feature of SPC, resulting from the used material of silicone elastomer PDMS, a relatively cheap material, which possesses high gas solubility and permeability. When the fabricated chips are evacuated, a negative pressure environment is formed due to the gas solubility of PDMS, which can service as a self-priming power to let the sample solutions be sucked into each reaction chamber and sequentially the biocompatible oil to seal and separate each filled chamber. Thus, in realizing dNAD, thousands of independent microwells can be created automatically, avoiding the external control system, which is superior to IFC cdNAD. Currently, SPC cdPCR and dINAAs [SPC chip-based dLAMP (cdLAMP) and SPC chip-based dIMSA (cdIMSA)] have been developed by our laboratory [30, 31, 69, 70].

3.3.2.1. SPC cdPCR

The chip shown in **Figure 7** is composed of three PDMS layers, two glass coverslips, and a waterproof layer [30]. The three PDMS layers include an inlet and outlet layer, a microwell array layer, and a blank layer. In the microwell array layer, a total of 5120 reaction microwells (150 μm in width, 150 μm in length, and 250 μm in height) are equally distributed in four separate panels. Each microwell contains down to 5 nL solution. The inlet and outlet layers have four 0.5-mm holes and four 2.5-mm holes in diameter punched as injection ports and suction chambers when aligning to the outlet of the microwell array layer, respectively. For mechanical stability, the blank layer coats the microwell array layer with the waterproof layer embedded. The waterproof layer is made of low permeability fluorosilane polymer, which beneficially prevents the evaporation during the step of denaturing template DNA at 95°C in PCR. One of the glass coverslips with plasma pretreated are used to seal the microwell array, and the other one is pressed on the upper surface of the SPC chip for mechanical stability at the end of microchip operation.

MSL techniques are used to fabricate the SPC chip. The chip patterns are designed by a software of CorelDRAW X4 and printed on transparency films using a high-resolution printer to create

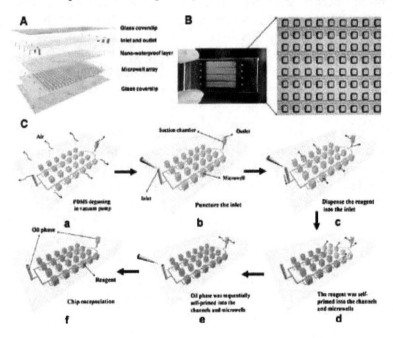

Figure 7. (A) Schematic diagram of the layered device structure of the SPC chip. (B) Photograph of the prototype SPC cdPCR device. The size of the chip is 50×24×4 mm. (C) Principle and operation procedure of the SPC microfluidic device. The red cuboids (150×150×250 μm) stand for the microwells.

the masks of channels and microwells. The photoresist material used are SU8 serials, which are a high-contrast, epoxy-based negative photoresist. Several 4-inch silicon wafers are adopted as the mold substrate. The PDMS to replicate the SPC chip is the silicone elastomer PDMS, which is composed of PDMS base (A) and catalyst (B) at certain ratios. Because dPCR is an endpoint detection, the reaction components including PCR buffer, primers, labeled probes, and templates have to be mixed before loading into the chip. Each diluted template in the mixture is individually injected into the three panels of the chip, allowing the samples to be compartmentalized completely and each chamber contains down to 5 nL solution. A Maestro Ex In-Vivo Imaging System (CRI Maestro, USA) is used to capture the fluorescent image of the microchip after dPCR. As a new generation of microfluidic chips, SPC cdPCR has been successfully applied to the absolute quantification of β-actin DNAs and the lung cancer-related genes.

3.3.2.2. SPC chip-based dINAAs (cdINAAs)

Although PCR is widely adopted and used as a standard analytical technique in molecular diagnosis, it is remarkably confined when applied to field and POCT due to the facts that it requires nonportable thermocycling facilities, its process of obtaining results is cumbersome, and the whole amplification takes 2 h or more. Also, SPC cdPCR confronts the same defects. Accordingly, SPC cdLAMP and SPC cdIMSA are established by our laboratory. As simple and easy world-to-chip fluidic devices, SPC cdINAAs have the great potential in POCT for the developing countries.

Similar to SPC cdPCR, SPC cdLAMP is also the completely valve-free and SPC device (**Figure 8**) [31]. It is also made mainly of PDMS and fabricated by MSL techniques. In size, the SPC chip used for dLAMP is the same as the dPCR-used chip; however, in composition, it does not contain a waterproof layer because of the absence of the DNA denaturing step in LAMP. For the microwell array PDMS layer of dLAMP-used SPC chip, a total of 4800 microwells (150 μm in width, 150 μm in length, and 300 μm in height) are fabricated and they are also equally distributed into four panels (each contains 1200 chambers), and the interval for two closed chambers is 150 μm. The big difference from the dPCR-used SPC is that the rectangular chambers are located vertically on the main channels and the branch channels link to chamber without orthogonal turning points. On performance, the SPC cdLAMP can precisely calculate the absolute DNA concentration. To conduct the data acquisition and analysis of SPC cdLAMP, the Maestro Ex In-Vivo Imaging System is employed. However, the imaging system is too cumbersome and expensive to allow the e POCT, especially in the less developed regions. Herein, an easy-to-use and cost-efficient smartphone-based dLAMP POCT device platform is also established by our laboratory.

SPC cdIMSA is an updated version of SPC cdLAMP, in which the LAMP is replaced by IMSA and a mixed dye is used to establish a label-free and sensitive dual-fluorescence detection for on-chip IMSA [70, 71]. The used SPC chip for dIMSA is the same as that for dPCR without any modifications. The SPC cdIMSA with the mixed dye for the detection of hepatitis B virus (HBV) is conducted. In contrast, the mixed dye indicating two different colors makes it easy to count the positive chamber number by visual inspection regardless of the cutoff values. Also, there

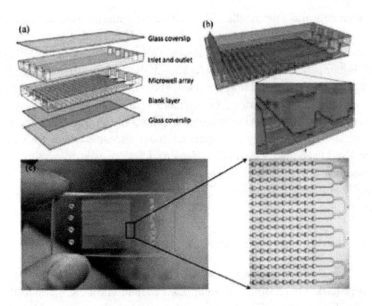

Figure 8. (a) Schematic diagram of the SPC chip for cdLAMP. (b) Schematic of the whole microchip and the enlarged schematic diagram of the part of the chip, with insets showing the array and microwell geometries. It contains four separate panels, each of which has an individual inlet and outlet. The blue lines (8×50 μm) are the flow channel. The red spots (150×150×300 μm) stand for the microwells. Each microwell was partitioned by oil. (c) Photograph of the prototype SPC cdLAMP device.

is a linear response of the counted positive chamber number to the dilution ratios of templates. Moreover, the dual fluorescence is capable of indicating more positive chamber number in contrast to single fluorescence by SYBR Green I. The mixed dye-loaded SPC cdIMSA possesses the advantage of enlarging the color changes over other currently used dyes or indicators. Similarly, the mixed dye-based dual-fluorescence detection has a potential in the POCT application of SPC cdLAMP and other SPC chip-based dINAAs.

3.3.3. LTNP cdNAD

As another POCT-oriented LOC device, LTNP chip resembling SPC chip also uses the gas solubility of PDMS to load solution into chambers (**Figure 9**) [29]. However, the measure of evacuating the LTNP chip is different from the SPC chip. For the former, the gas solubility and permeability of the chip itself and another PDMS layer coating on the chip services simultaneously as the real-time power source of evacuating with a syringe filter. For the latter, preevacuating by a vacuum pump is indispensable, which calls for robust and efficient approaches of sealing and packaging chips [30].

The LTNP cdPCR has been already exploited [29]. The chip consists of three parts, which are a lamina-chip layer (LCL), a vaporproof layer (VPL), and a syringe filter-like microfluidic device (μfilter) with helical channels. The μfilter has two parts. One is designed for generating the LTNP through pulling the plug of the syringe connected at one end of helical channels. The helical channel in the μfilter is 200 μm in width and 40 μm in depth. The other is used for sample and oil loading via the punched ports (0.5 mm in diameter) aligning at the ones in the LCL. The area diameter of the region of helical channel in the two µfilters is 20 mm, covering entirely the area of the whole chambers in the LCL. Sandwiched by the two parts of the µfilter, LCL contains 650 chambers in a square array. The chamber is cylindrical with 200 µm in diameter and 200 µm in depth, and the distance between two closed chambers is also 200 µm. VPL on the LCL is also a thin circular layer with chambers at 100 mm in diameter, and its area is 17.0 mm in diameter, entirely mulching the LCL.

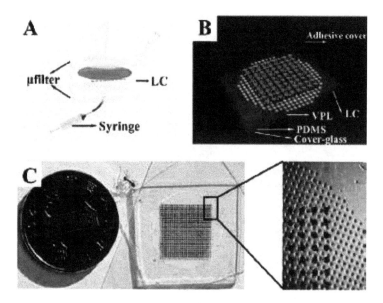

Figure 9. (A) Schematic diagram of the LTNP chip. (B) Schematic of the reagent-loaded chip. It contains reagent-loaded lamina chip (red), water-loaded VPL (blue), PDMS on the coverglass, and optical adhesive cover. (C) Digital image of the prototype LTNP cdPCR device.

Currently, LTNP cdPCR has been used for the detection of keratin 19 in A549 lung carcinoma cells [29]. Additionally, the integrated LTNP chip with NA enrichment, isolation, and digital detection functions has been developed and successfully applied to detect bovine DNA in ovine meat for food adulteration detection [72, 73].

3.3.4. Other cdNADs

Gansen et al. described a self-digitization (SD) cdLAMP device shown in **Figure 10** [32]. In this device, the reagents were partitioned into the microchambers based on an inherent fluidic phenomenon that the interplay between fluidic forces and interfacial tension could cause the self-dispersion of an income aqueous fluid into an array of chambers prefilled with an immiscible fluid. Therefore, the sample loading could be realized with manual or automated syringe pumps or external air pressure, removing the hydraulic valves or mechanical action. Less than 2 µL sample was used for the accurate quantification of relative and absolute DNA concentrations. To improve the efficiency of partitioning samples, a new generation of SD chip was invented with close to 100% efficiency in 2013 [74]. In 2014, based on the SD chip, digital RT-PCR was developed to absolutely quantify mRNA from single cells [75]. Due to the simplicity and robustness of the SD chip, the SD cdNAD is an inexpensive and easy-to-operate digital detection device.

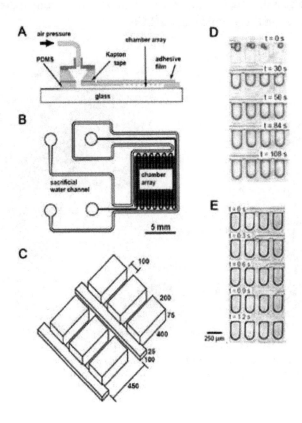

Figure 10. Design of the SD cdLAMP device. (A) Schematic diagram of the individual components of a fully assembled chip. Air pressure was delivered via a removable adapter, which was connected to an external pressure source. (B) Layout of the microfluidic network. A dense array of rectangular side chambers was connected to a thin main channel. The whole array was surrounded by a separate water reservoir to saturate the PDMS during incubation at 65°C. Scale bar, 5 mm. (C) Geometry of the side chamber array and main channel. All dimensions are in micrometers. (D) Sequential images showing the initial filling of the side-chamber array with aqueous solution. (E) Sequence of images showing the SD of aqueous sample in the side chambers.

In 2010, Sundberg et al. designed a spinning disk platform to achieve dPCR [25]. The disk was an inexpensive and disposable plastic disk-like chip. Differing from other approaches, 1000 nL microwells were generated by passive compartmentalization through centrifugation, and the volume of each well was 33 nL in average. The whole process, including disk loading, thermocycling, and fluorescent imaging, only costs less than 35 min. However, this kind of cdNAD performed some defects, such as the tedious plastic disk manufacturing and the probable NA adsorption to the disk.

4. Analysis and evaluation of cdNAD

This part focuses on the analysis and evaluation of cdNAD. The contents involve four aspects, which are the dehydration of microreaction, the dynamic range, the response of cdNAD, and the precision of cdNAD.

4.1. Dehydration of microreaction

In PCR, dehydration is attributable to the repeated denaturing steps, whereas, for isothermal NA amplification, reacting at a consistent temperature (e.g., 63°C) for at least 1 h results in dehydration. Also, the influence of dehydration is different according to the different material-based chips for cdNAD.

As we know, the microfluidic PDMS chips suffer from dehydration heavily. Actually, the dehydration degree is determined by the thickness of the PDMS layer between the top of the chamber and the waterproof layer. If v_f is the total volume fraction of reaction reagent, v_f is defined by the function of

$$v_f = \frac{A_{chamber} \cdot h_{chamber}}{A_{PDMS} \cdot h_{PDMS}}$$

where A and h refer to the designed area and height, respectively. Using the formula of $C_{sat\_25°C} \times P_{vap\_70°C}/P_{vap\_25°C}$, a saturated concentration of water vapor in PDMS at 70°C ($C_{sat\_70°C}$) can be calculated as 400 mol/m³. Then, the maximum fractional loss of water (fl_{max}) from the reaction chambers is defined by

where ρ_{water} is the density of water and M_{water} is the molar mass of water.

4.2. Dynamic range

When high concentration targets are loaded, the chip panel can be completely saturated, whereas, for low targets, the chip panel still possesses the capability of realizing digital detection. Therefore, the theoretical dynamic range is determined by the high concentrations that make the chip completely saturated.

In this situation, the occurrence of an empty chamber is a small probability event, and due to statistical independence between the chamber number and the total empty chamber number, the event can be modeled as a random Poisson process. Then, the occurrence probability of t empty chamber number, $P(n=t, \lambda)$, is defined by $P(n=t, \lambda) = {\lambda^t e^{-\lambda}}/{t!}$. In the function, λ equals to the mean number of empty chambers in panels. For each empty chamber in each panel, namely, a chamber containing 0 molecule, the probability $P(n=0,\lambda)=e^{-\lambda}=e^{-m/N}$, where λ is the ratio of the loaded molecule number (m) to the chamber numbers (N). Therefore, λ is equal to the product of the probability $P(n=0,\lambda)$ and the N chamber numbers, namely, $\lambda=Ne^{-m/N}$. Finally,

$P(n=t, \lambda)=P(n=t, Ne^{-m/N})=(Ne^{-m/N})^t \Big/ e^{-Ne^{-m/N}} \Big| t!$. If λ is more than 10, the probability of the chip panel is completely saturated, $P(n=0,\lambda)=e-10$, which is less than 10-4 and can be as the acceptable failure rate. Then, $\lambda=N e^{-m/N}>10$, and the maximum loaded molecular number (mmax) can be calculated.

4.3. Response of cdNAD

According to the distribution of molecules across the chip, the Poisson distribution is adapted to calculate the original concentration of target stock solution.

For each chip panel, the probability (P) of having the number (n) of NA molecules per chamber is $P(n,\lambda)=(\lambda^n e^{-\lambda})/n!$, where λ is the ratio of loaded molecule number to chamber number of each panel. For λ, it is equal to the average molecule number per chamber, which can be expressed by the equation of $\lambda=C_0 X_{dil} V$, where C_0 is the target's original concentration, V is the volume of each chamber, and X_{dil} is the dilution factor of diluted targets used in the panel. If the stock solution is diluted in k fold, the $X_{dil}=1/k$, and particularly, the X_{dil} of stock solution is 1. When a chamber contains 0 molecule, the probability $P(n=0,\lambda)=e^{-\lambda}$. As for a chamber capturing one or more molecules, the probability $P(n\geq1,\lambda)=1-P(n=0,\lambda)=1-e^{-\lambda}$. After dPCR, a chamber with observed positive signal suggests that at least one target molecule is captured. Therefore, the ratio (f) of observed positive chamber number to chamber number of each panel, namely, the observed fraction of positive chambers, equals to the probability $P(n\geq1,\lambda)$, which is $f=P(n\geq1,\lambda)=1-e^{-\lambda}$. Then, $\lambda=ln(1-f)=-C_0 X_{dil} V$, and a linear variation relationship is exhibited in terms of the regression curve equation between $ln(1-f)$ and X_{dil}, because C_0 and V are constant values. Then, the target's original concentration C_0 can be calculated based on the slope of the curve ($-C_0 V$).

In addition, because the Poisson distribution is a particular case of a binomial distribution, the distribution of the positive chamber number (x) in each panel is classified as the binomial distribution. The probability distribution $P(x)$ is therefore influenced by the probability $p=P(n\geq1,\lambda)=1-P(n=0,\lambda)=1-e^{-\lambda}$. Based on the nature of binomial distribution, x is equal to the mean, namely, $x=Np=N(1-e^{-\lambda})$ (N, the total chamber number in each panel). Then, the average molecule number per chamber, $\lambda=ln\{N/(N-x)\}$, through which the loaded molecule number (N_e) for each panel can be estimated by the equation $N_e=N\lambda=Nln\{N/(N-x)\}$.

4.4. Precision of cdNAD

When the chamber number in each panel is very large, the Poisson distribution is approximate to normal distribution, which is also called Gaussian distribution. Use the parameters in the discussion of response and dynamic range above, and let y be a random variable representing the number of positive chambers that capture at least one molecule. Also, we can know that its mathematical expectation or mean μ is $N(1-e^{-\lambda})$, and its variance σ^2 is $Ne^{-\lambda}(1-e^{-\lambda})$. Then, the probability density function associated with y is as follows:

If the precision of cdNAD is defined as the minimum difference in concentration ($\Delta\lambda/\lambda$) that is reliably detected with more than 99% true positive and more than 99% true negative rate, this situation corresponds to 4.6σ separation in the mean (μ) for two Gaussian distributions. That is to say, in the case with small $\Delta\lambda$, $\Delta\mu/\Delta\lambda=4.6\sigma$. Then, $\partial[N(1-e^{-\lambda})]\Delta\lambda/\partial\lambda=Ne^{-\lambda}\Delta\lambda=4.6[Ne^{-\lambda}(1-e^{-\lambda})]^{1/2}$, and $\Delta\lambda/\lambda=4.6(e^{\lambda}-1)^{1/2}/(\lambda N^{1/2})$. For different total chamber numbers (N) in chip panel, the precision ($\Delta\lambda/\lambda$) to expected molecules per chamber (λ) could be plotted.

5. Application of dNAD

Compared to conventional NA detection methods, dNAD provides better sensitivity, better precision, higher tolerance, and the NA's absolute quantification. Until now, dNAD has been applied to a variety of research fields, including pathogen detection, food safety, clinical diagnosis (genetic instability estimation and early cancer), prenatal diagnosis, quantitative analysis of gene expression, and NGS library quantification. Although dPCR is still the most widespread type of dNAD, in resource-limited regions, it is confined due to the requirement of thermal cycling and robust temperature control. To overcome this awkwardness, dINAAs without thermal cycling are of great interest, which also enlarge the application of dNAD, especially for POCT. For this subchapter, the recent advances on application of dNAD will be narrated.

5.1. Pathogen detection

Honestly speaking, on pathogen detection, the advantages of most conventional NADs on both analytical sensitivity and specificity are inherited by the corresponding digital formats. In addition, dNAD has the capability of realizing NA's absolute quantification. Therefore, dNAD can be used as the ultrahigh sensitivity and more accurate method for viral load determination or bacterial quantification. Kelley et al. established a duplex ddPCR assay for high-precision methicillin-resistant *Staphylococcus aureus* (MRSA) analysis [76]. By assaying 397 clinical samples, a good agreement with the reference assay for both qPCR and ddPCR assays was indicated. Strain et al. demonstrated and estimated the application of ddPCR for a highly precise measurement of HIV DNA [77]. Total HIV DNA and episomal 2-LTR (long terminal repeat) circles in cells that were all isolated from infected patients were targeted. Compared to qPCR, ddPCR performed a significantly increased precision (5-fold for total HIV DNA and >20-fold for 2-LTR circles), making it an alternative for the measurement of HIV DNA from clinical specimens.

dNAD also enables the direct detection without NA extraction due to the ability of partitioning target NA and nontarget components (e.g., inhibitors) into different microwells. Pavšič et al. employed two dPCR platforms (QX100 ddPCR system from Bio-Rad Laboratories and the BioMark HD IFC cdPCR system from Fluidigm) for the direct quantification of two whole-virus materials of human cytomegalovirus (HCMV) without DNA extraction [78]. It was demonstrated that direct quantification by both dPCRs could provide repeatable measurements of viral DNA copy numbers, giving a closer agreement with the actual viral load than that with either dPCR or qPCR of extracted DNA.

Although dNAD certainly possesses the superiority of absolute quantitation, some reports fail to demonstrate the advantage, which calls for the requirement of optimization when establishing dNAD. For instance, Boizeau et al. used ddPCR to provide an absolute quantitation of HBV genome molecules [79]. However, the results indicated that qPCR assays remained more sensitive than ddPCR when used for low HBV DNA levels, suggesting that optimization of ddPCR was still necessary, especially on accurately differentiating the positive from negative in samples with very low levels of target DNA molecules.

5.2. Food safety

Food safety mainly refers to two parts: the minoring of foodborne pathogens and the detection of GMO. For example, human noroviruses (NoV) and hepatitis A virus (HAV) are two foodborne enteric viruses that have caused the vast majority of nonbacterial gastroenteritis or some fatal infectious hepatitis. Coudray-Meunier et al. conducted a comparative study of IFC RT-cdPCR and conventional RT-qPCR when quantifying the NoV and HAV from lettuce and water samples, proving that the IFC RT-cdPCR assay was more tolerant to inhibitory substances from lettuce samples [80]. Also, the IFC RT-cdPCR may be useful for standardizing the quantification of enteric viruses in bottled water and lettuce samples. Fu et al. used the BioMark HD system equipped with a 48.770 digital array to develop cdPCR for GMO detection without pretreatment [81]. The CaMV35s promoter and the NOS terminator were selected as the targets, and nine events of GMOs (MON810, MON863, TC1507, MIR604, MIR162, GA21, T25, NK603, and Bt176) were collected to determine the specificity. The results showed that the cdPCR could achieve a discrimination of down to 0.1%, lower than the labeling threshold level of the EU, allowing highly sensitive, specific, and stable GMO screening detection. Dalmira et al. developed a duplex ddPCR assay to characterize the certified reference materials (CRMs) in terms of T-*nos*/*hmg* copy number ratio in maize [82]. After optimization using a central composite design, the duplex ddPCR method realized the absolute limit of detection (LOD) and limit of quantification (LOQ) of 11 and 23 copy number T-*nos*, namely, relative LOD of 0.034% and relative LOQ of 0.08%, respectively. The dynamic range of T-*nos*/*hmg* ratio ranged from 0.08% to 100%. The results indicated that the duplex ddPCR assay was useful for characterizing CRM candidates on T-*nos*/*hmg* ratio.

The bias caused by reliance on quantitative standards may have an impact on the results of qPCR, which is not beneficial for water monitoring and microbial source identification. Therefore, Cao et al. employed a duplex ddPCR to simultaneously quantify *Enterococcus* spp. and the human fecal-associated HF183 marker for evaluating the water quality [83]. The results demonstrated that ddPCR performed greater tolerance of inhibition than qPCR, with one to two orders of magnitude higher at inhibitor concentrations. Also, ddPCR brought about remarkably improved precision, although a lower upper LOQ than qPCR was indicated.

For food safety, it is indispensable to identify and quantify the meat products for unveiling species fraud and product mislabeling during food processing. Tian et al. used LTNP cdPCR to detect bovine meat in ovine meat [72]. Floren et al. successfully applied two-step ddPCR for precise quantification of cattle, horse, and pig in processed meat product, with the nuclear *F2* gene targeted [84].

5.3. Clinical diagnosis

At present, clinical diagnosis including genetic instability estimation and early cancer is also the main application field for dNAD. It has been clearly confirmed that the genetic instability of human cells is one of the causes of cancer, including somatic mutation, allelic imbalance, loss of heterozygosity, CNVs, and single nucleotide variations (SNVs). Accordingly, how to discriminate the rare mutant gene from abundant normal NAs attracts great concerns of researchers. Interestingly, it is also the question that initiates the concept formation of dNAD.

As *de novo* CNV may be caused by reprogramming somatic cells into induced pluripotent stem cells, Abyzov et al. used qPCR and ddPCR to detect and estimate this phenomenon. The results showed that, in parental fibroblasts, at least half of the CNVs are indicated as low-frequency somatic genomic variants [85]. Boettger et al. successfully applied a ddPCR approach to analyze father-mother-offspring trios from HapMap at specific sites within region 1 on the investigation of inference of complex CNV and single nucleotide polymorphism (SNP) haplotypes at the human 17q21.31 locus [86].

Detecting biomarkers associated with tumor formation, development, and drug evaluations on gene expression is the useful approach for early cancer diagnosis. Up to now, various biomarkers have been identified, and dNAD reflects the unique advantage of absolute quantitation of their gene expression [87]. Zhu et al. applied SPC cdPCR to the detection of three lung cancer-related genes (PLAU, ENO2, and PLAT). cdPCR yielded comparable results to qPCR, illustrating that the established platform had the ability of realizing absolute quantitation for gene expression [30]. Floren et al. employed ddPCR to detect the BRAF-V600E and V600K mutations in melanoma circulating tumor with high sensitivity [84]. The study demonstrated that ddPCR performed 200-fold increased sensitivity than competitive allele-specific PCR (castPCR), giving an LOD of 0.0005% when combined with whole-genome amplification (WGA). Through noninvasive analysis of circulating free plasma DNA, Geven-sleben et al. determined the presence of oncogenic amplification by developing a plasma DNA ddPCR assay targeting HER2 [88]. In the independent validation cohort, ddPCR could reach a positive and negative predictive value of 70% and 92%, respectively. The results suggested that ddPCR had the potential to the analysis of any locus amplified in cancer, not only in metastatic breast cancer. Beaver et al. also employed a ddPCR assay for the detection of circulating plasma tumor DNA (ptDNA) in patients with early-stage cancer [89]. A total of 30 tumors were first analyzed by Sanger sequencing for common PIK3CA mutations, and ddPCR was then used to analyze their extracted DNA for the same mutations. This ddPCR-based accurate mutation detection platform was demonstrated to be of great use for early-stage breast cancer.

As a new generation of cancer biomarkers, abnormally DNA methylation in the gene's regulatory regions can affect identical residues that may cause the cancer. The analysis of methylated genes therefore becomes more and more important in cancer research. For example, Li et al. developed a sensitive bead-based ddPCR for the quantification of DNA methylation, which could detect down to one methylated DNA molecule in approximately 5000 unmethylated DNA molecules from plasma or fecal samples [20]. Weisenberger et al. developed an improved DNA methylation detection with single-molecule high-resolution based on IFC cdPCR and successfully identified the breast cancer-specific hypermethylation phenomenon in the CpG islands of RUNX3, CLDN5, and FOXE1 [90].

5.4. Prenatal diagnosis

Prenatal diagnosis is usually carried out by invasive or noninvasive approaches [91]. The invasive method (e.g., amniocentesis) refers to inserting needles into the uterus, which is time-consuming (several weeks required) and risky to the fetus [92]. In contrast, noninvasive method seems to be more rapid and safer. Currently, noninvasive prenatal diagnosis mainly involves ultrasonography of the womb or detecting cell-free fetal DNA (cffDNA) in maternal serum and plasma. The latter is also termed as noninvasive prenatal testing (NIPT). Appling dNAD to NIPT is the emerging noninvasive approach with high sensitivity and high precision [93–97]. Gu et al. employed the QX100 ddPCR system from Bio-Rad Laboratories to detect cffDNA for the risk of methylmalonic acidemia, confirming that ddPCR is a cost-effective and noninvasive prenatal method when diagnosing known mutations associated with Mendelian disorders [98]. Pornprasert and Prasing implemented a ddPCR for the deletion of $\alpha(0)$-thalassemia Southeast Asian (SEA)-type deletion [99]. The study showed that ddPCR might be an alternative technology to routine clinical diagnosis. Barrett et al. adopted 12×765 digital array chips to establish an IFC cdPCR to analyze the cffDNA for NIPT of sickle cell anemia [94]. The results suggested that the built IFC cdPCR is a useful method to determine the genotype of fetuses at risk for sickle cell anemia. Meantime, the report also illustrated that it was essential to optimize the fractional fetal DNA concentration.

Because most cffDNA fragments were approximately 200 bp in size and, in the early gestation, the cffDNA occupies low percentage (mostly <10%) in maternal plasma, efficient methods for the extraction of cffDNA are in great need. Holmberg et al. thereby estimated two commercial platforms to extract cffDNA: the Akonni Biosystems TruTip technology and the Circulating Nucleic Acids Kit from Qiagen [100]. Determined by QX100 ddPCR system and qPCR, the extracted products from two platforms performed similar results.

5.5. NGS library quantification

As known to all, the establishment of NGS libraries still leans mainly on manual bench-top procedures, which is slow and inefficient [101]. To solve it, Kim et al. invented an automated digital microfluidic LOC-based sample preparation for the NGS. Compared to the conventional methods, digital microfluidic LOC platform is cost-efficient and has high throughout [102]. Similarly, Thaitrong et al. integrated a droplet-based microfluidic LOC system with a unit of capillary-based reagent delivery and the quantitative CE module to develop an automated quality-control platform for NGS [103]. Besides, White et al. confirmed that dPCR is able to provide sensitive and absolute calibration for NGS, enabling direct sequencing without titration runs and with sufficient precision [11]. Fu et al. developed a picoliter-scaled droplet digital WGA (ddWGA) platform for realizing uniform and accurate single-cell sequencing, bringing about significantly improved amplification evenness and accuracy for the simultaneous detection of CNVs and SNVs in single-cell level [104]. Weisenberger et al. applied the Fluidigm BioMark Digital Array to establish digital bisulfite genomic DNA sequencing with high resolution and high sensitivity [105]. The results showed that IFC cdPCR was a fast and reliable method for the single-molecule-scaled detection of DNA methylation information.

6. Conclusion and future prospects

Undoubtedly, there is growing interest in dNAD, because it allows the more precise NA quantification, the higher discrimination of rare NA mutants, and the more reproducible and less susceptible to inhibitors than the traditional NA methods. Consequently, dNAD has full potential to influence the development of biology research, clinical diagnosis, the safety of food and environment, and other research fields.

To enhance the impact of this promising technique and push it towards clinical application, the MIQE for dNAD (dMIQE) was also published [106]. Based on dMIQE, the experimental protocols are standardized, the efficient utilization of resources is maximized, and the data are adequately assisted. However, the promising technique of dNAD still confronts some shortcomings. On one hand, although the development of microfluidic LOC offers a lot of dNAD device platforms, these devices perform low functional integration, and the supporting detection approaches lean primarily on real-time fluorescence scanning or the endpoint analysis of CCD camera-captured images, which, to some extent, adds the real cost and also has impacts on the true detection accuracy. Therefore, in the future, it will become a general trend that dNAD devices are highly integrated with multiple functions including cell or single-cell capture, cell lysis, and NA enrichment and purification, employing more advanced supporting detection technology. Particularly, for ddNAD, the strategy of droplet generation is one of the developing directions. For instance, Tanaka et al. currently created a hands-off autonomous preparation method of monodisperse emulsion droplets using a degassed PDMS chip [107]. Jeong et al. used a specially designed three-dimensional monolithic elastomer device to create a kiloscale droplet generation [108]. According to the snap-off mechanism, Barkley et al. also invented a novel technique to generate monodisperse droplets [109].

On the other hand, dNAD is actually the digital version of NAD, thereby possessing the same disadvantages (e.g., bias or nonspecific amplification) as most sequence-based NA amplification methods. Based on this point, how to improve or guarantee the NA amplification fidelity in microreactors is one of the future prospects of dNAD. It should be noted that the optimized reaction conditions for NA amplification in microreactors might be different from those in bulk state, meaning that the optimization of reaction system is indispensable. Furthermore, the precision of dNAD is greatly influenced by the number and size of microchambers (for cdNAD) or droplet (for ddNAD), which in turn challenges the fabrication of microfluidic chip and the uniformity of partitioning. Accordingly, we also anticipate that, in the future, there will be some novel strategies developed to realize digital detection not only based on partitioning reagents.

Certainly, the application of dNAD will be enlarged by combining with other molecular assays, especially for single-cell analysis and single-cell genomic sequencing. For example, dNAD can combine with proximity ligation or PEAs to achieve single-molecule protein biomarker detection. Additionally, in the future, ongoing comparison tests of dNAD and qPCR will roundly prove the detection superiority of dNAD in many research fields. Also, dNAD will become the promising POCT-oriented research area for the ambitious plan of precision medicine.

Conclusively, dNAD based on microfluidic LOC devices will continue to provide further opportunities for determining NA molecules, protein molecules, and other biomolecules towards deep analysis with high sensitivity and precision.

Author details

Xiong Ding[1,2] and Ying Mu[1*]

*Address all correspondence to: muying@zju.edu.cn

1 Research Center for Analytical Instrumentation, Institute of Cyber-Systems and Control, State Key Laboratory of Industrial Control Technology, Zhejiang University, Hangzhou, P.R. China

2 College of Life Sciences, Zhejiang University, Hangzhou, P.R. China

References

[1] Biomarkers SMAPN, Gilad S, Meiri E, Yogev Y, Benjamin S, Lebanony D, et al. Serum microRNAs are promising novel biomarkers. PLoS ONE. 2008;3(9):e3148.

[2] Blakely WF. Nucleic acid molecular biomarkers for diagnostic biodosimetry applications: use of the fluorogenic 5′-nuclease polymerase chain reaction assay. Military Medicine. 2002;167(2):16–19.

[3] Chen X, Ba Y, Ma L, Cai X, Yin Y, Wang K, et al. Characterization of microRNAs in serum: a novel class of biomarkers for diagnosis of cancer and other diseases. Cell Research. 2008;18(10):997–1006.

[4] Schwarzenbach H, Hoon DS, Pantel K. Cell-free nucleic acids as biomarkers in cancer patients. Nature Reviews Cancer. 2011;11(6):426–437.

[5] Tost J. DNA methylation: an introduction to the biology and the disease-associated changes of a promising biomarker. Molecular Biotechnology. 2010;44(1):71–81.

[6] Chen Y, Perkins M, Teixeira L, Cave M, Eisenach K. Comparison of the ABI 7700 system (TaqMan) and competitive PCR for quantification of IS6110 DNA in sputum during treatment of tuberculosis. Journal of Clinical Microbiology. 1998;36(7):1964–1968.

[7] Tobal K, Newton J, Macheta M, Chang J, Morgenstern G, Evans P, et al. Molecular quantitation of minimal residual disease in acute myeloid leukemia with t(8;21) can identify patients in durable remission and predict clinical relapse. Blood. 2000;95(3): 815–819.

[8] Gack MU, Kirchhofer A, Shin YC, Inn K-S, Liang C, Cui S, et al. Roles of RIG-I N-terminal tandem CARD and splice variant in TRIM25-mediated antiviral signal transduction. Proceedings of the National Academy of Sciences. 2008;105(43):16743–16748.

[9] Neely LA, Patel S, Garver J, Gallo M, Hackett M, McLaughlin S, et al. A single-molecule method for the quantitation of microRNA gene expression. Nature Methods. 2006;3(1): 41–46.

[10] Jarvius J, Melin J, Göransson J, Stenberg J, Fredriksson S, Gonzalez-Rey C, et al. Digital quantification using amplified single-molecule detection. Nature Methods. 2006;3(9): 725–727.

[11] White RA, Blainey PC, Fan HC, Quake SR. Digital PCR provides sensitive and absolute calibration for high throughput sequencing. BMC Genomics. 2009;10(1):116.

[12] Shiroguchi K, Jia TZ, Sims PA, Xie XS. Digital RNA sequencing minimizes sequence-dependent bias and amplification noise with optimized single-molecule barcodes. Proceedings of the National Academy of Sciences. 2012;109(4):1347–1352.

[13] Shuga J, Zeng Y, Novak R, Lan Q, Tang X, Rothman N, et al. Single molecule quanti-tation and sequencing of rare translocations using microfluidic nested digital PCR. Nucleic Acids Research. 2013:41(16):e159.

[14] Pfaffl MW. A new mathematical model for relative quantification in real-time RT-PCR. Nucleic Acids Research. 2001;29(9):e45–e45.

[15] Čikoš Š, Bukovská A, Koppel J. Relative quantification of mRNA: comparison of methods currently used for real-time PCR data analysis. BMC Molecular Biology. 2007;8(1):113.

[16] Whelan JA, Russell NB, Whelan MA. A method for the absolute quantification of cDNA using real-time PCR. Journal of Immunological Methods. 2003;278(1):261–269.

[17] Bustin SA. Absolute quantification of mRNA using real-time reverse transcription polymerase chain reaction assays. Journal of Molecular Endocrinology. 2000;25(2):169–193.

[18] Schmittgen TD, Livak KJ. Analyzing real-time PCR data by the comparative CT method. Nature Protocols. 2008;3(6):1101–1108.

[19] Zachar V, Thomas RA, Goustin AS. Absolute quantification of target DNA: a simple competitive PCR for efficient analysis of multiple samples. Nucleic Acids Research. 1993;21(8):2017.

[20] Li M, Chen W-d, Papadopoulos N, Goodman SN, Bjerregaard NC, Laurberg S, et al. Sensitive digital quantification of DNA methylation in clinical samples. Nature Biotechnology. 2009;27(9):858–863.

[21] Fu GK, Wilhelmy J, Stern D, Fan HC, Fodor SP. Digital encoding of cellular mRNAs enabling precise and absolute gene expression measurement by single-molecule counting. Analytical Chemistry. 2014;86(6):2867–2870.

[22] Whale AS, Cowen S, Foy CA, Huggett JF. Methods for applying accurate digital PCR analysis on low copy DNA samples. PLoS ONE. 2013;8(3):e58177.

[23] Vogelstein B, Kinzler KW. Digital PCR. Proceedings of the National Academy of Sciences. 1999;96(16):9236–9241.

[24] Men Y, Fu Y, Chen Z, Sims PA, Greenleaf WJ, Huang Y. Digital polymerase chain reaction in an array of femtoliter polydimethylsiloxane microreactors. Analytical Chemistry. 2012;84(10):4262–4266.

[25] Sundberg SO, Wittwer CT, Gao C, Gale BK. Spinning disk platform for microfluidic digital polymerase chain reaction. Analytical Chemistry. 2010;82(4):1546–1550.

[26] Shen F, Du W, Kreutz JE, Fok A, Ismagilov RF. Digital PCR on a SlipChip. Lab on a Chip. 2010;10(20):2666–2672.

[27] Fan HC, Blumenfeld YJ, El-Sayed YY, Chueh J, Quake SR. Microfluidic digital PCR enables rapid prenatal diagnosis of fetal aneuploidy. American Journal of Obstetrics and Gynecology. 2009;200(5):e541–543, e547.

[28] Pinheiro LB, Coleman VA, Hindson CM, Herrmann J, Hindson BJ, Bhat S, et al. Evaluation of a droplet digital polymerase chain reaction format for DNA copy number quantification. Analytical Chemistry. 2011;84(2):1003–1011.

[29] Tian Q, Song Q, Xu Y, Zhu Q, Yu B, Jin W, et al. A localized temporary negative pressure assisted microfluidic device for detecting keratin 19 in A549 lung carcinoma cells with digital PCR. Analytical Methods. 2015;7(5):2006–2011.

[30] Zhu Q, Qiu L, Yu B, Xu Y, Gao Y, Pan T, et al. Digital PCR on an integrated self-priming compartmentalization chip. Lab on a Chip. 2014;14(6):1176–1185.

[31] Zhu Q, Gao Y, Yu B, Ren H, Qiu L, Han S, et al. Self-priming compartmentalization digital LAMP for point-of-care. Lab on a Chip. 2012;12(22):4755–4763.

[32] Gansen A, Herrick AM, Dimov IK, Lee LP, Chiu DT. Digital LAMP in a sample self-digitization (SD) chip. Lab on a Chip. 2012;12(12):2247–2254.

[33] Blainey PC, Quake SR. Digital MDA for enumeration of total nucleic acid contamination. Nucleic Acids Research. 2011;39(4):e19–e19.

[34] Shen F, Davydova EK, Du W, Kreutz JE, Piepenburg O, Ismagilov RF. Digital isothermal quantification of nucleic acids via simultaneous chemical initiation of recombinase polymerase amplification reactions on SlipChip. Analytical Chemistry. 2011;83(9): 3533–3540.

[35] Mazutis L, Araghi AF, Miller OJ, Baret J-C, Frenz L, Janoshazi A, et al. Droplet-based microfluidic systems for high-throughput single DNA molecule isothermal amplification and analysis. Analytical Chemistry. 2009;81(12):4813–4821.

[36] Rane TD, Chen L, Zec HC, Wang T-H. Microfluidic continuous flow digital loop-mediated isothermal amplification (LAMP). Lab on a Chip. 2015;15(3):776–782.

[37] Schuler F, Schwemmer F, Trotter M, Wadle S, Zengerle R, von Stetten F, et al. Centrifugal step emulsification applied for absolute quantification of nucleic acids by digital droplet RPA. Lab on a Chip. 2015;15(13):2759–2766

[38] Konry T, Smolina I, Yarmush JM, Irimia D, Yarmush ML. Ultrasensitive detection of low-abundance surface-marker protein using isothermal rolling circle amplification in a microfluidic nanoliter platform. Small. 2011;7(3):395–400.

[39] Pohl G, Shih I-M. Principle and applications of digital PCR. Expert Rev Mol Diagn. 2004;4(1):41–47.

[40] Barrett AN, Chitty LS. Developing noninvasive diagnosis for single-gene disorders: the role of digital PCR. Methods in Molecular Biology. 2014;1160:215–228.

[41] Lo YD, Lun FM, Chan KA, Tsui NB, Chong KC, Lau TK, et al. Digital PCR for the molecular detection of fetal chromosomal aneuploidy. Proceedings of the National Academy of Sciences. 2007;104(32):13116–13121.

[42] Baker M. Digital PCR hits its stride. Nature Methods. 2012;9(6):541–544.

[43] Morrison T, Hurley J, Garcia J, Yoder K, Katz A, Roberts D, et al. Nanoliter high throughput quantitative PCR. Nucleic Acids Research. 2006;34(18):e123–e123.

[44] Marx V. PCR: paths to sensitivity. Nature Methods. 2014;11(3):241–245.

[45] Majumdar N, Wessel T, Marks J. Digital PCR modeling for maximal sensitivity, dynamic range and measurement precision. PLoS ONE. 2015;10(3):e0118833.

[46] Conte D, Verri C, Borzi C, Suatoni P, Pastorino U, Sozzi G, et al. Novel method to detect microRNAs using chip-based QuantStudio 3D digital PCR. BMC Genomics. 2015;16(1): 1.

[47] Klančnik A, Toplak N, Kovač M, Marquis H, Jeršek B. Quantification of *Listeria monocytogenes* cells with digital PCR and their biofilm cells with real-time PCR. Journal of Microbiological Methods. 2015;118:37–41.

[48]   Kinz E, Leiherer A, Lang A, Drexel H, Muendlein A. Accurate quantitation of JAK2
       V617F allele burden by array-based digital PCR. International Journal of Laboratory
       Hematology. 2015;37(2):217–224.

[49]   Kaitu'u-Lino T, Hastie R, Cannon P, Lee S, Stock O, Hannan NJ, et al. Stability of
       absolute copy number of housekeeping genes in preeclamptic and normal placentas,
       as measured by digital PCR. Placenta. 2014;35(12):1106–1109.

[50]   Schweitzer P, Harris A, Mandelman D, Jackson S, Cifuentes F, Degoricija L. Precise
       quantification of next generation sequencing Ion Torrent™ and Illumina Libraries
       using the QuantStudio™ 3D Digital PCR Platform. Journal of Biomolecular Techni-
       ques. 2014;25(Suppl):S15.

[51]   Leng X, Zhang W, Wang C, Cui L, Yang CJ. Agarose droplet microfluidics for highly
       parallel and efficient single molecule emulsion PCR. Lab on a Chip. 2010;10(21):2841–
       2843.

[52]   Kennedy, Suzanne. PCR troubleshooting and optimization: the essential guide.
       Horizon Scientific Press, 2011.

[53]   Williams R, Peisajovich SG, Miller OJ, Magdassi S, Tawfik DS, Griffiths AD. Amplifi-
       cation of complex gene libraries by emulsion PCR. Nature Methods. 2006;3(7):545–550.

[54]   Tavvfik DS, Griffiths AD. Man-made cell-like compartments for molecular evolution.
       Nature Biotechnology. 1998;16(7):652–656

[55]   Margulies M, Egholm M, Altman WE, Attiya S, Bader JS, Bemben LA, et al. Genome
       sequencing in microfabricated high-density picolitre reactors. Nature. 2005;437(7057):
       376–380.

[56]   Selck DA, Karymov MA, Sun B, Ismagilov RF. Increased robustness of single-molecule
       counting with microfluidics, digital isothermal amplification, and a mobile phone
       versus real-time kinetic measurements. Analytical Chemistry. 2013;85(22):11129–
       11136.

[57]   Beer NR, Wheeler EK, Lee-Houghton L, Watkins N, Nasarabadi S, Hebert N, et al. On-
       chip single-copy real-time reverse-transcription PCR in isolated picoliter droplets.
       Analytical Chemistry. 2008;80(6):1854–1858.

[58]   Zhong Q, Bhattacharya S, Kotsopoulos S, Olson J, Taly V, Griffiths AD, et al. Multiplex
       digital PCR: breaking the one target per color barrier of quantitative PCR. Lab on a
       Chip. 2011;11(13):2167–2174.

[59]   Hindson BJ, Ness KD, Masquelier DA, Belgrader P, Heredia NJ, Makarewicz AJ, et al.
       High-throughput droplet digital PCR system for absolute quantitation of DNA copy
       number. Analytical Chemistry. 2011;83(22):8604–8610.

[60]   Liu W, Chen D, Du W, Nichols KP, Ismagilov RF. SlipChip for immunoassays in
       nanoliter volumes. Analytical Chemistry. 2010;82(8):3276–3282.

[61] Li L, Ismagilov RF. Protein crystallization using microfluidic technologies based on valves, droplets, and SlipChip. Biophysics. 2010;39:139–158

[62] Shen F, Du W, Davydova EK, Karymov MA, Pandey J, Ismagilov RF. Nanoliter multiplex PCR arrays on a SlipChip. Analytical Chemistry. 2010;82(11):4606–4612.

[63] Sun B, Shen F, McCalla SE, Kreutz JE, Karymov MA, Ismagilov RF. Mechanistic evaluation of the pros and cons of digital RT-LAMP for HIV-1 viral load quantification on a microfluidic device and improved efficiency via a two-step digital protocol. Analytical Chemistry. 2013;85(3):1540–1546.

[64] Xia Y, Liu Z, Yan S, Yin F, Feng X, Liu B-F. Identifying multiple bacterial pathogens by loop-mediated isothermal amplification on a rotate & react slipchip. Sensors and Actuators B: Chemical. 2016;228:491–499.

[65] Unger MA, Chou H-P, Thorsen T, Scherer A, Quake SR. Monolithic microfabricated valves and pumps by multilayer soft lithography. Science. 2000;288(5463):113–116.

[66] Ottesen EA, Hong JW, Quake SR, Leadbetter JR. Microfluidic digital PCR enables multigene analysis of individual environmental bacteria. Science. 2006;314(5804):1464–1467.

[67] Heyries KA, Tropini C, VanInsberghe M, Doolin C, Petriv I, Singhal A, et al. Megapixel digital PCR. Nature Methods. 2011;8(8):649–651.

[68] Xu L, Lee H, Jetta D, Oh KW. Vacuum-driven power-free microfluidics utilizing the gas solubility or permeability of polydimethylsiloxane (PDMS). Lab on a Chip. 2015;15(20):3962–3979.

[69] Song Q, Gao Y, Zhu Q, Tian Q, Yu B, Song B, et al. A nanoliter self-priming compartmentalization chip for point-of-care digital PCR analysis. Biomedical Microdevices. 2015;17(3):1–8.

[70] Ding X, Wu W, Zhu Q, Zhang T, Jin W, Mu Y. Mixed-dye-based label-free and sensitive dual fluorescence for the product detection of nucleic acid isothermal multiple-self-matching-initiated amplification. Analytical Chemistry. 2015;87(20):10306–10314.

[71] Ding X, Nie K, Shi L, Zhang Y, Guan L, Zhang D, et al. Improved detection limit in rapid detection of human enterovirus 71 and coxsackievirus A16 by a novel reverse transcription-isothermal multiple-self-matching-initiated amplification assay. Journal of Clinical Microbiology. 2014;52(6):1862–1870.

[72] Tian Q, Mu Y, Xu Y, Song Q, Yu B, Ma C, et al. An integrated microfluidic system for bovine DNA purification and digital PCR detection. Analytical Biochemistry. 2015;491:55–57.

[73] Tian Q, Yu B, Mu Y, Xu Y, Ma C, Zhang T, et al. An integrated temporary negative pressure assisted microfluidic chip for DNA isolation and digital PCR detection. RSC Advances. 2015;5(100):81889–81896.

[74] Schneider T, Yen GS, Thompson AM, Burnham DR, Chiu DT. Self-digitization of samples into a high-density microfluidic bottom-well array. Analytical Chemistry. 2013;85(21):10417–10423.

[75] Thompson AM, Gansen A, Paguirigan AL, Kreutz JE, Radich JP, Chiu DT. Self-digitization microfluidic chip for absolute quantification of mRNA in single cells. Analytical Chemistry. 2014;86(24):12308–12314.

[76] Kelley K, Cosman A, Belgrader P, Chapman B, Sullivan DC. Detection of methicillin-resistant *Staphylococcus aureus* by a duplex droplet digital PCR assay. Journal of Clinical Microbiology. 2013;51(7):2033–2039.

[77] Strain MC, Lada SM, Luong T, Rought SE, Gianella S, Terry VH, et al. Highly precise measurement of HIV DNA by droplet digital PCR. PLoS ONE. 2013;8(4):e55943.

[78] Pavšič J, Žel J, Milavec M. Digital PCR for direct quantification of viruses without DNA extraction. Analytical and Bioanalytical Chemistry. 2016;408(1):67–75.

[79] Boizeau L, Laperche S, Désiré N, Jourdain C, Thibault V, Servant-Delmas A. Could droplet digital PCR be used instead of real-time PCR for quantitative detection of the hepatitis B virus genome in plasma? Journal of Clinical Microbiology. 2014;52(9):3497–3498.

[80] Coudray-Meunier C, Fraisse A, Martin-Latil S, Guillier L, Delannoy S, Fach P, et al. A comparative study of digital RT-PCR and RT-qPCR for quantification of hepatitis A virus and norovirus in lettuce and water samples. International Journal of Food Microbiology. 2015;201:17–26.

[81] Fu W, Zhu P, Wang C, Huang K, Du Z, Tian W, et al. A highly sensitive and specific method for the screening detection of genetically modified organisms based on digital PCR without pretreatment. Scientific Reports. 2015; 5:12715. doi:10.1038/srep12715.

[82] Dalmira Fl-Ud, Melina Pr-U, José-Benigno VT, Josefina Ln-Fl, Raymundo Ga-E, Abraham A-S. Development, optimization, and evaluation of a duplex droplet digital PCR assay to quantify the T-nos/hmg copy number ratio in genetically modified maize. Analytical Chemistry. 2015;88(1):812–819.

[83] Cao Y, Raith MR, Griffith JF. Droplet digital PCR for simultaneous quantification of general and human-associated fecal indicators for water quality assessment. Water Research. 2015;70:337–349.

[84] Floren C, Wiedemann I, Brenig B, Schütz E, Beck J. Species identification and quantification in meat and meat products using droplet digital PCR (ddPCR). Food Chemistry. 2015;173:1054–1058.

[85] Abyzov A, Mariani J, Palejev D, Zhang Y, Haney MS, Tomasini L, et al. Somatic copy number mosaicism in human skin revealed by induced pluripotent stem cells. Nature. 2012;492(7429):438–442.

[86] Boettger LM, Handsaker RE, Zody MC, McCarroll SA. Structural haplotypes and recent evolution of the human 17q21. 31 region. Nature Genetics. 2012;44(8):881–885.

[87] Day E, Dear PH, McCaughan F. Digital PCR strategies in the development and analysis of molecular biomarkers for personalized medicine. Methods. 2013;59(1):101–107.

[88] Gevensleben H, Garcia-Murillas I, Graeser MK, Schiavon G, Osin P, Parton M, et al. Noninvasive detection of HER2 amplification with plasma DNA digital PCR. Clinical Cancer Research. 2013;19(12):3276–3284.

[89] Beaver JA, Jelovac D, Balukrishna S, Cochran R, Croessmann S, Zabransky D, et al. Detection of cancer DNA in plasma of early stage breast cancer patients. Clinical Cancer Research. 20(10):2643–50.

[90] Weisenberger DJ, Liang G. Contributions of DNA methylation aberrancies in shaping the cancer epigenome. Translational Cancer Research. 2015;4(3):219–234.

[91] Collins S, Impey L. Prenatal diagnosis: types and techniques. Early Human Development. 2012;88(1):3–8.

[92] Young C, von Dadelszen P, Alfirevic Z. Instruments for chorionic villus sampling for prenatal diagnosis (review). Cochrane Database of Systematic Reviews. 2013;1:CD000114.

[93] Debrand E, Lykoudi A, Bradshaw E, Allen SK. A non-invasive droplet digital PCR (ddPCR) assay to detect paternal CFTR mutations in the cell-free fetal DNA (cffDNA) of three pregnancies at risk of cystic fibrosis via compound heterozygosity. PLoS ONE. 2015;10(11):e0142729.

[94] Barrett AN, McDonnell TC, Chan KA, Chitty LS. Digital PCR analysis of maternal plasma for noninvasive detection of sickle cell anemia. Clinical Chemistry. 2012;58(6): 1026–1032.

[95] Svobodová I, Pazourková E, Hořínek A, Novotná M, Calda P, Korabečná M. Performance of droplet digital PCR in non-invasive fetal RHD genotyping—comparison with a routine real-time PCR based approach. PLoS ONE. 2015;10(11):e0142572.

[96] Jin S, Lin XM, Law H, Kwek KY, Yeo GS, Ding C. Further improvement in quantifying male fetal DNA in maternal plasma. Clinical Chemistry. 2012;58(2):465–468.

[97] Kantak C, Chang C-P, Wong CC, Mahyuddin A, Choolani M, Rahman A. Lab-on-a-Chip technology: impacting non-invasive prenatal diagnostics (NIPD) through miniaturisation. Lab on a Chip. 2014;14(5):841–854.

[98] Gu W, Koh W, Blumenfeld YJ, El-Sayed YY, Hudgins L, Hintz SR, et al. Noninvasive prenatal diagnosis in a fetus at risk for methylmalonic acidemia. Genetics in Medicine. 2014;16(7):564–567.

[99] Pornprasert S, Prasing W. Detection of α(0)-thalassemia South-East Asian-type deletion by droplet digital PCR. European Journal of Haematology. 2014;92(3):244–248.

[100]   Holmberg RC, Gindlesperger A, Stokes T, Lopez D, Hyman L, Freed M, et al. Akonni
        TruTip® and Qiagen® methods for extraction of fetal circulating DNA—evaluation by
        real-time and digital PCR. PLoS ONE. 2013;8(8):e73068.

[101]   Voelkerding KV, Dames SA, Durtschi JD. Next-generation sequencing: from basic
        research to diagnostics. Clinical Chemistry. 2009;55(4):641–658.

[102]   Kim H, Jebrail MJ, Sinha A, Bent ZW, Solberg OD, Williams KP, et al. A microfluidic
        DNA library preparation platform for next-generation sequencing. PLoS ONE.
        2013;8(7):e68988.

[103]   Thaitrong N, Kim H, Renzi RF, Bartsch MS, Meagher RJ, Patel KD. Quality control of
        next-generation sequencing library through an integrative digital microfluidic plat-
        form. Electrophoresis. 2012;33(23):3506–3513.

[104]   Fu Y, Li C, Lu S, Zhou W, Tang F, Xie XS, et al. Uniform and accurate single-cell
        sequencing based on emulsion whole-genome amplification. Proceedings of the
        National Academy of Sciences. 2015;112(38):11923–11928.

[105]   Weisenberger DJ, Trinh BN, Campan M, Sharma S, Long TI, Ananthnarayan S, et al.
        DNA methylation analysis by digital bisulfite genomic sequencing and digital Meth-
        yLight. Nucleic Acids Research. 2008;36(14):4689–4698.

[106]   Huggett JF, Foy CA, Benes V, EMSLie K, Garson JA, Haynes R, et al. The digital MIQE
        guidelines: minimum information for publication of quantitative digital PCR experi-
        ments. Clinical Chemistry. 2013;59(6):892–902.

[107]   Tanaka H, Yamamoto S, Nakamura A, Nakashoji Y, Okura N, Nakamoto N, et al.
        Hands-off preparation of monodisperse emulsion droplets using a poly (dimethylsi-
        loxane) microfluidic chip for droplet digital PCR. Analytical Chemistry. 2015;87(8):
        4134–4143.

[108]   Jeong H-H, Yelleswarapu VR, Yadavali S, Issadore D, Lee D. Kilo-scale droplet
        generation in three-dimensional monolithic elastomer device (3D MED). Lab on a Chip.
        2015;15(23):4387–4392.

[109]   Barkley S, Weeks ER, Dalnoki-Veress K. Snap-off production of monodisperse droplets.
        European Physical Journal E. 2015;38(12):1–4.

Permissions

All chapters in this book were first published by InTech Open; hereby published with permission under the Creative Commons Attribution License or equivalent. Every chapter published in this book has been scrutinized by our experts. Their significance has been extensively debated. The topics covered herein carry significant findings which will fuel the growth of the discipline. They may even be implemented as practical applications or may be referred to as a beginning point for another development.

The contributors of this book come from diverse backgrounds, making this book a truly international effort. This book will bring forth new frontiers with its revolutionizing research information and detailed analysis of the nascent developments around the world.

We would like to thank all the contributing authors for lending their expertise to make the book truly unique. They have played a crucial role in the development of this book. Without their invaluable contributions this book wouldn't have been possible. They have made vital efforts to compile up to date information on the varied aspects of this subject to make this book a valuable addition to the collection of many professionals and students.

This book was conceptualized with the vision of imparting up-to-date information and advanced data in this field. To ensure the same, a matchless editorial board was set up. Every individual on the board went through rigorous rounds of assessment to prove their worth. After which they invested a large part of their time researching and compiling the most relevant data for our readers.

The editorial board has been involved in producing this book since its inception. They have spent rigorous hours researching and exploring the diverse topics which have resulted in the successful publishing of this book. They have passed on their knowledge of decades through this book. To expedite this challenging task, the publisher supported the team at every step. A small team of assistant editors was also appointed to further simplify the editing procedure and attain best results for the readers.

Apart from the editorial board, the designing team has also invested a significant amount of their time in understanding the subject and creating the most relevant covers. They scrutinized every image to scout for the most suitable representation of the subject and create an appropriate cover for the book.

The publishing team has been an ardent support to the editorial, designing and production team. Their endless efforts to recruit the best for this project, has resulted in the accomplishment of this book. They are a veteran in the field of academics and their pool of knowledge is as vast as their experience in printing. Their expertise and guidance has proved useful at every step. Their uncompromising quality standards have made this book an exceptional effort. Their encouragement from time to time has been an inspiration for everyone.

The publisher and the editorial board hope that this book will prove to be a valuable piece of knowledge for researchers, students, practitioners and scholars across the globe.

List of Contributors

Veronica Iacovacci, Gioia Lucarini, Leonardo Ricotti and Arianna Menciassi
The BioRobotics Institute, Scuola Superiore Sant'Anna, Pisa, Italy

Louis WY Liu, Qingfeng Zhang and Yifan Chen
Department of Electrical Electronics Engineering, South University of Science Technology of China, Shenzhen, China

Mahdi Mohammadi, David J Kinahan and Jens Ducrée
School of Physical Sciences, National Centre for Sensor Research, Dublin City University (DCU), Dublin, Ireland

Yehya H. Ghallab
Department of Biomedical Engineering, Helwan University, Cairo, Egypt
Centre of Nanoelectronics and Devices (CND) at Zewail City of Science and Technology/ American University in Cairo, Cairo, Egypt

Yehea Ismail
Centre of Nanoelectronics and Devices (CND) at Zewail City of Science and Technology/ American University in Cairo, Cairo, Egypt

Ferenc Ender
Department of Electron Devices, Budapest University of Technology and Economics, Budapest, Hungary

Diána Weiser
Department of Organic Chemistry and Technology, Budapest University of Technology and Economics, Budapest, Hungary

László Poppe
Department of Organic Chemistry and Technology, Budapest University of Technology and Economics, Budapest, Hungary
SynBiocat LLC, Lázár deák, Budapest, Hungary

Nicola Massimiliano Martucci, Nunzia Migliaccio, Immacolata Ruggiero, Annalisa Lamberti and Paolo Arcari
Department of Molecular Medicine and Medical Biotechnology, University of Naples, Federico II, Naples, Italy

Ilaria Rea, Monica Terracciano, Luca De Stefano and Ivo Rendina
Institute for Microelectronics and Microsystems of Naples, National Research Council, Naples, Italy

Preeti Nigam Joshi
Organic Chemistry Division, National Chemical Laboratory, Pune, India

Xiong Ding
Research Center for Analytical Instrumentation, Institute of Cyber-Systems and Control, State Key Laboratory of Industrial Control Technology, Zhejiang University, Hangzhou, P.R. China
College of Life Sciences, Zhejiang University, Hangzhou, P.R. China

Ying Mu
Research Center for Analytical Instrumentation, Institute of Cyber-Systems and Control, State Key Laboratory of Industrial Control Technology, Zhejiang University, Hangzhou, P.R. China

Index

Printed in the USA
CPSIA information can be obtained
at www.ICGtesting.com
JSHW051621061123
51533JS00005B/53

9 781639 877027